Ecological Studies, Vol. 95

Analysis and Synthesis

Guest Editor

R.A.A. Oldeman (Eds.)

Ecological Studies

Volume 75
The Grazing Land Ecosystems of the African Sahel (1989)
H.N. Le Houérou

Volume 76
Vascular Plants as Epiphytes: Evolution and Ecophysiology (1989)
U. Lüttge (Ed.)

Volume 77
Air Pollution and Forest Decline: A Study of Spruce (*Picea abies*) on Acid Soils (1989)
E.-D. Schulze, O.L. Lange, and R. Oren (Eds.)

Volume 78
Agroecology: Researching the Ecological Basis for Sustainable Agriculture (1990)
S.R. Gliessman (Ed.)

Volume 79
Remote Sensing of Biosphere Functioning (1990)
R.J. Hobbs and H.A. Mooney (Eds.)

Volume 80
Plant Biology of the Basin and Range (1990)
B. Osmond, G.M. Hidy, and L. Pitelka (Eds.)

Volume 81
Nitrogen in Terrestrial Ecosystem: Questions of Productivity, Vegetational Changes, and Ecosystem Stability (1990)
C.O. Tamm (Eds.)

Volume 82
Quantitative Methods in Landscape Ecology: the Analysis and Interpretation of Landscape Heterogeneity (1990)
M.G. Turner and R.H. Gardner (Eds.)

Volume 83
The Rivers of Florida (1990)
R.J. Livingston (Ed.)

Volume 84
Fire in the Tropical Biota: Ecosystem Processes and Global Challenges (1990)
J.G. Goldammer (Ed.)

Volume 85
The Mosaic- Cycle Concept of Ecosystems (1991)
H. Remmert (Ed.)

Volume 86
Ecological Heterogeneity (1991)
J. Kolasa and S.T.A. Pickett (Eds.)

Volume 87
Horses and Grasses: The Nutritional Ecology of Equids and Their Impact on the Camargue (1992)
P. Duncan

Volume 88
Pinnipeds and E1 Nino: Responses to Enviromental Stress (1992)
F. Trillmich and K.A. Ono (Eds.)

Volume 89
Plantago: A Multidisciplinary Study (1992)
P.J.C. Kuiper and M. Bos (Eds.)

Volume 90
Biogeochemistry of a Subalpine Ecosystem: Loch Vale Watershed (1992)
J. Baron (Ed.)

Volume 91
Atmospheric Deposition and Forest Nutrient Cycling (1992)
D.W. Johnson and S.E. Lindberg (Eds.)

Volume 92
Landscape Boundaries (1992)
A.J. Hansen and F. di Castri (Eds.)

Volume 93
Fire in South African Mountain Fynbos (1992)
B.W. van Wilgen, D.M. Richardson, F.J. Kruger and H. J. van Hensbergen (Eds.)

Volume 94
The Ecology of Aquatic Hyphomycetes (1992)
F. Bärlocher (Ed.)

Francis Kahn Jean-Jacques de Granville

Palms in Forest Ecosystems of Amazonia

With 64 Figures and 67 Tables

Springer-Verlag
Berlin Heidelberg New York
London Paris Tokyo
Hong Kong Barcelona
Budapest

Dr. Francis Kahn
Institut Français de Recherche Scientifique
pour le Développement en Coopération (ORSTOM),
Apartado postal 18-1209
Lima 18, Peru

Dr. Jean-Jacques de Granville
Institut Français de Recherche Scientifique
pour le Développement en Coopération (ORSTOM),
BP 165
97323 Cayenne cédex, France

ISBN 3-540-54399-6 Springer-Verlag Berlin Heidelberg NewYork
ISBN 0-387-54399-6 Springer-Verlag NewYork Berlin Heidelberg

Library of Congress Cataloging-in-Publication Data
Kahn, Francis. Palms in forest ecosystems of Amazonia / Francis Kahn, Jean Jacques de Granville. p. cm. – (Ecological studies; vol. 95). Includes bibliographical references and index.
ISBN 3-540-54399-6. – ISBN 0-387-54399-6
1. Palms – Amazon River Region. 2. Palms – Amazon River Region – Ecology. 3. Forest ecology – Amazon River Region. I. Granville, Jean-Jacques de. II. Title. III. Series: Ecological studies; v. 95. QK495.P17K34 1992 584′.5′09811 – dc20 91-44729 CIP

Typesetting and printing: Best-set Typesetter Ltd., Hong Kong
31/3145-5 4 3 2 1 0 – Printed on acid-free paper

Contents

Preface

Palms are tropical miracles. Heinrich Heine, the German poet, stated "*Unter den Palmen wandert man nicht ungestraft*", i.e., one does not wander unpunished under the palms. It was Professor H.C.D. de Wit who taught me this in the late 1950s, and it is a pleasure to forward this message to the next generation in such an appropriate book.

Both authors, as I know them, will bear the punishment of the palms. They will never be without palm nostalgia if and when living somewhere outside this world's tropical and subtropical palm belt. Palm nostalgia goes further than palms alone. It concerns the landscape, the short but splendid sunsets and last, but not least, the tropical people. Their elegance of living, structured in subtler ways than managers will ever understand, their laughter which may be a more decisive weapon against the troubles besetting the tropics than mere economics, and their unique life force erupting on festive as well as sad occasions under the palms will always remain with those who wandered beneath these trees. I know. I was there.

Palms are mythological beings, and to a great extent the present book proves the myths to be true. This has to do with the fact that it concerns Amazonia and the Guianas, the home of palm civilizations. In a dry, all-but severely scientific style, the uses of palm trees are described in a later chapter. But for those having witnessed the making of blowpipes from *Iriartella* stems, or the weaving of hammocks from the leaf initia of 200 palms, the wonder is there, behind the text clad in a modest black suit.

This severity, which is in no way a flaw in scientific writings, is compensated by the beautiful drawings and sketches, combining precision with the projection of images one remembers having seen. There is more, and better, quantitative work in such scale drawings than in many a page covered by printouts and formulas. This has a profound reason, which was very recently discovered, i.e., the interplay among many regular, sometimes even linear processes, producing not a new regularity, but an unpredictable chaos. The average values, frequencies, and sizes calculated for plants, plants species, or plant communities with statistical means, only partly explain the architecture and dynamics of biological or ecological systems.

We do not have fine-turned methods available to calculate the tendencies and structures beyond statistics, to calculate the unpredictable in a seemingly determinate and severe world of mathematical natural laws. Hence, drawings remain the only tools to see the links and/or missing links in biological structure and dynamics. It has to be remembered that we know the flora and fauna from Ancient Egypt only because some artists painted plants and animals on walls. Descriptions like those in the Bible show that no clear image of plants can be captured in words; drawings are needed. The present book on Amazonian palms will show many human generations pictorially what palms grew in Amazonia in the second half of our century, how they grew and how they looked. It has good chances to become a classic. Besides the beautiful pictures and drawings, the book will be useful because of its big key and descriptions of Amazonian palms.

This is by no means exaggerated, because like Richard's *The Tropical Rain Forest* in 1952, the present book resumes knowledge of a group that has not yet been so completely treated since Martius. The descriptions of so many interactions with other animal and plant components of the different rain forests of Amazonia are an example of this. The number of critical remarks, which will certainly follow, is to be seen as proof that an unexpectedly great number of scientists have not only bought, but read this book. This is an exceptional feat, nowadays.

I should wish this book in many, many hands, so that many people start to know, or know better, those magnificent and elegant giants among monocotyledons, but still only Davids next to the dicotyledonous Goliaths.

Roelof A.A. Oldeman

Acknowledgments

We are particularly indebted to Jean Louis Guillaumet, who leads the ORSTOM research unit on the diversity and functioning of forest ecosystems. This book originates from the many discussions we had walking in the forest of central Amazonia in 1980. Francis Hallé and Roelof A.A. Oldeman have always been good supporters of these activities and helped us in many ways. Moreover, many of the illustrations were drawn in the field by F. Hallé in Peruvian Amazonia and many stimulating suggestions and productive criticism have been made by R.A.A. Oldeman who reviewed the text and agreed to write a Preface to this book. We also thank Anthony B. Anderson, Michael J. Balick, Anders Barfod, Rodrigo Bernal, Patrick Blanc, Finn Borchsenius, Christian Feuillet, Gloria Galeano, Dennis Johnson, Kember Mejia, Christine Padoch, and Gérard Tavakilian for the discussion on palms and related topics, J. Dransfield, who helped with the identification of specimens, and Guy Couturier and Andrew Henderson, who agreed to read the manuscript and made many corrections and useful comments. We express our acknowledgments to all those responsible for the Institutes with which we collaborated: the late Jean Marie Michotte (ORSTOM-Cayenne), Roger Beuzeville, José López Parodi, Jaime Moro (Instituto de Investigación de la Amazonía Peruana, Iquitos), Emma Cerrate and Gerardo Lamas (Museo de Historia Natural-Universidad Nacional Mayor San Marcos, Lima), and Herbert O. Schubart (Instituto Nacional de Pesquisas da Amazônia, Manaus). We were assisted in many ways by the staff of the Botany Department in Cayenne: George Cremers, curator of the herbarium, and Michel Hoff, responsible for the Aublet computerized data bank. We would like to compliment P. Buirette for all printing and enlargements of black and white photographs. We express our special thanks to Farana Moussa and Annick Altuna, who spent a good deal of time typing and reviewing the text, and to Joe McCann for his helpful assistance on the English manuscript.

Photographs were taken by J.J. de Granville (Figs. 11, 12, 15, 20, 32, 33, 36, 37, 42, 43, 44, 55), and F. Kahn (Figs. 23, 57, 58, 61, 62). Drawings of palms are by J.P. Gasc (Fig. 18b), J.J. de Granville (Figs. 7, 9, 10, 16, 19, 25, 26, 28, 29, 35, 45, 46), F. Hallé (Figs. 2, 3, 8, 13, 14, 21, 24, 31, 39, 41), F. Kahn (Figs. 6, 17, 18a,d, 22), R. Pacheco (Figs. 4, 5, 6, 18c), and M. Ocrospoma (Figs. 17, 27, 40, 59, 60, 63).

Our gratitude goes to the editors and authors who kindly granted permission to use the following material: Biotropica (Figs. 47, 50, 51; Tables 18, 39, 40, 42), Experientia (Figs. 34b, 52, 53, 54; Tables 17, 43A,B, 44, 45, 47); Bulletin de l'IFEA (Fig. 63); Forest Ecology and Management (Tables 31, 32, 33, 34, 36); Principes (Fig. 64; Tables 20, 21, 28, and the keys for palm identification): J.P. Gasc (Fig. 18c); Bois et forêt des tropiques (Fig. 15): SEPANGUY (Figs. 7, 9, 10, 16, 19, 25, 35, 45, 46); P. Sist (Tables 12A,B, 60).

Introduction

Palms are like a watchword in Amazonia. Knowing and speaking about palms is your key to acceptance everywhere. Everyone knows about them; they are present in everyone's daily life. They are the most characteristic components of the Amazonian landscapes, in the field as well as in paintings on the walls of Iquitos airport and restaurants, or in the famous Theatre of Manaus. Even though the primary forest is being destroyed by man, palms remain the last arborescent plants to remind us of the luxuriant jungle.

Traveling throughout the Amazon basin, most of the earlier naturalist-explorers were attracted by palms. D'Orbigny, Wallace, Martius, Drude, Barbosa Rodrigues, Poiteau, Spruce, and Trail left us basic studies on palm taxonomy with numerous notes on their ecology and usage by the natives. From 1928 to 1956, Burret, a botanist at Berlin-Dahlem Museum, described many new species and published taxonomic treatments of many genera, from material collected by Tessmann, Ule, and others in Amazonia. Moore's expedition in Peru took place in 1960; a new era was beginning, which would culminate with the appearance of Uhl and Dransfield's *Genera Palmarum* (1987), a classification of palms based on his work. The 1960s were also marked by Wessels Boer's study of indigenous palms of Surinam (1965), as well as by his revision of the genus *Geonoma* (1968). More recently, palms were intensively collected by Prance, Balick, and Henderson in Brazil; Foster, Gentry, and Smith in Peru; Balslev in Ecuador; and many other botanists, all of whom have contributed to our knowledge of Amazonian palm flora. The studies of Amazonian palms have been essentially oriented upon taxonomy and ethnobotany, and little quantitative data on the diversity or the density of palms in the forests were provided. Most forestry and botanical inventories that included plants with >10 cm DBH (diameter at 1.3 m in height) took into account palms with trunks (Black et al. 1950; Pires et al. 1953; Lechthaler 1956; Prance et al. 1976; Gentry 1982; Boom 1986; Campbell et al. 1986; Balslev et al. 1987), but they did not consider the majority of smaller palms.

How many palm species are there in Amazonian forest ecosystems? Is palm species richness equal throughout the Amazon basin? Do palms develop a similar range of life forms in the forest ecosystems of a restricted area, or within the same forest ecosystem in different regions of the basin? What about palm density and life forms? These are the questions that this

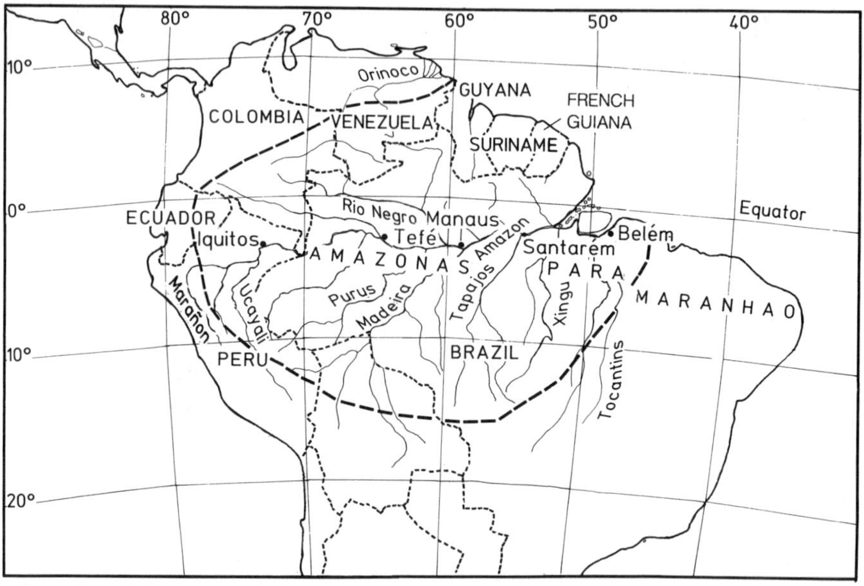

Fig. 1. Amazonia

book aims at answering. The definition of the status of palms in the forest ecosystems in the Amazon basin, i.e., their diversity and the place they occupy in each ecosystem, will give the basic knowledge which must be the foundation for any discussion on their evolution and adaptation.

One problem was to define Amazonia. A hydrographic definition restricted to the Amazon basin, i.e., the Amazon River and all its tributaries, would not take into account the phytogeographic coherency. In fact, we have adopted Spruce's concept (1871, p. 69), "the great Amazonian forest, extending northward to the cataracts of the Orinoco, in Venezuela, southward far into the centre of Brazil, and westward almost to the very crest of the Andes . . ." (Fig. 1). Amazonia is the vast continuum of uplands and wetlands covered in tropical rain forests, which extend on both sides of the Amazon from the eastern Andean slopes and valleys, around 2000–2500 m in altitude, to the Atlantic coast, and include some low shrubby vegetation and savannahs. *Ceroxylon* appears beyond the western limit. The southern and northern limits are not so clearly defined because of the overlapping of forests, shrubby vegetation, and savannahs; at the southern limit of the basin occur *Copernicia* and *Trithrinax*. The Guianas are part of such a set. How could one deal with Amazonian palms without including the Guianan species? Most of them are widely distributed throughout the basin.

Many publications present the characteristics of the Amazon basin as regards its geography and geomorphology (Sternberg 1950, 1955, 1957; Journaux 1975; Tricart 1977), hydrology and climates (Klinge 1965; Sioli 1968, 1984; Villa Nova et al. 1976; Salati and Ribeiro 1979; Leopoldo et al.

1982a,b; Salati and Marques 1984; Salati and Vose 1984; Junk and Furch 1985; Salati 1985; Leopoldo et al. 1987), and soils (Sombroek 1966; Chauvel 1982; Jordan 1985; Chauvel et al. 1987). The functioning of Amazonian forest ecosystems has also been particularly analyzed (Fittkau and Klinge 1973; Klinge 1977a,b; Herrera et al. 1978; Jordan 1978; Stark and Jordan 1978; Franken et al. 1979; Irmler and Furch 1980; Jordan et al. 1980; Jordan and Heuveldop 1981; Walker and Franken 1983; Schubart et al. 1984; Uhl and Jordan 1984; Herrera 1985; Luizão and Schubart 1987). Also the vegetation types have been described by many authors (Takeuchi 1960, 1962; Aubréville 1961; de Granville 1978; Pires 1978; Prance 1978, 1979; Encarnación 1985; Guillaumet 1987; Daly and Prance, 1989).

The reader will not find any sophisticated analyses to corroborate ecological theories. The numerous tables containing quantitative data are included only to describe the variability of the life forms developed by palms and the structure of the palm communities in the forests. The aim of this book is to draw a picture of the palm situation in the Amazonian forests: we cannot help adding . . . before they completely disappear. Some may find this position rather pessimistic. We consider it realistic. Several of the forests studied here have since been destroyed. This is the case of the splendid terra firme forests dominated by the Brazil nut tree, which have been permanently flooded after damming of the lower Tocantins River valley (huge hydroelectric dam of Tucuruí). The deforested areas are increasing perilously (Spears 1979; Fearnside 1982; Houghton et al. 1991). Colonization in Rondônia has cleared between 37200 and 37900 km^2 by 14 July 1988 (Stone et al. 1991); the opening of roads and uncontrolled shifting-cultivation; extensive and most often unproductive pastures; vast, superficial mining projects such as those of Serra dos Carajas; the much-debated Jari project (Fearnside and Rankin 1979, 1982); and coca cultivation and the subsequent deforestation of the eastern Andean piedmont: all of these are sadly famous examples of forest destruction. Many authors have pointed out the jeopardy of Amazonian forests (Gómez-Pompa et al. 1972; Fearnside 1984; Prance 1986; Uhl and Saldarriaga 1987). Forseeable effects of deforestation on climate and hydrology (Gentry and López Parodi 1980; Nordin and Meade 1982) and social cost (Moran 1984) have been analyzed and summed up by Sioli (1991, p. 131): definite losses of the nutrients; surface erosion; decrease in water percolation; reduction of evapotranspiration and of the annual pluviosity; more unequal distribution of the rains and more unstable regime of the rivers; increase in sediments; higher air and soil temperatures; increase of the CO_2-content of the atmosphere; extermination of the richest and most diverse genestock of the earth; extermination of indigenous human cultures. As P. Nogueira-Neto wrote in the preface of Prance and Lovejoy's book (1985), "If we don't learn quickly how to benefit from the rational use of forest products, without destroying the forest, conservation will be a lost cause in Amazonia. We have little time left to act before it is too late."

Numerous data presented in this book have been published separately. They are gathered together here and completed by hitherto unpublished data on the variability of life forms, new inventories of species, productivity of several species. The whole is original. The attempt has been made, in particular, to describe the major trends of adaptive radiation of palms in Amazonia as well as to underline their role in the functioning of the forest ecosystems. Many observations will have to be completed. Hence, if this work does no more than motivate new studies in order to better understand forest ecosystems, its aims will have been partially accomplished.

Chapter 1 presents the subfamilies and genera of Amazonian palms and the general lines of their geographic distribution. An estimated number of species is given for Amazonia according to the latest taxonomic studies. Most genera are likely to be represented by only one or a few species which are highly variable in morphology. Consequently, these have been described as several or many different species. Moreover, the most diversified palm genus, *Bactris*, needs a serious reassessment. Therefore, the number of species given per genus is provisional. The life forms of the Amazonian palms are defined, characterized from quantitative morphological data, and illustrated in Chapter 2. Palms are considered as structural components of forest ecosystems in Chapter 3. The characteristics of palm communities in the forest ecosystems are exposed: richness of species, density, diversity in life forms, and vertical distribution of palms. Their variation through the Amazon basin within each ecosystem, as well as their differences from one ecosystem to another, are analyzed. Chapter 4 discusses palm distribution as a function of soil drainage, topography, forest architecture and dynamics, i.e., water and light. Life forms integrate morphological and functional features of the plant. With reference to the diversity of life forms in the ecosystem on the one hand and their ecological range on the other, comments will be made upon the major trends of the adaptive radiation of Palmae in Amazonia, and adaptive value of some palm structures will be discussed in Chapter 5. The importance of palms in ecosystem functioning is pointed out in Chapter 6: leaf and fruit productivity, relationship with animals, several features in relation to life forms of palms. Chapter 7 deals with palms and forest management. The knowledge of the place occupied by palms in the diversified forest ecosystems and the estimate of their economic potential lead to a definition of some priorities for swamp forest management.

A supplement offers a practical tool for dealing with Amazonian palms: a key for the identification of genera using vegetative characters applied to seedlings, juveniles, and sterile adults; an elementary description of each genus based on the characteristics of Amazonian species; and the most common vernacular names. This supplement will allow the reader to become familiar with Amazonian palms in the field, and should help him to start new studies on palms.

1 Amazonian Palm Flora – General Considerations

1.1 Richness of Amazonian Palm Flora

Since there has been no recent reassessment of many genera (Uhl and Dransfield 1987), it is difficult to give the real number of species within a genus, or, consequently, the total number of species occurring in Amazonia. This is likely to be lower in the largest genera, such as *Bactris* and *Astrocaryum*, where the related species, which present a high morphological variability, are considered as one. This attitude was assumed by Wessels Boer (1968) with *Geonoma*, Skov and Balslev (1989) with *Hyospathe*, and Henderson (1990) with *Iriartea*. On the other hand, the discovery of new genera and species for Amazonia is likely to continue for as long as the whole basin area has not been thoroughly collected. Since 1987, four new species have been named and others are to be described (Borchsenius, de Granville, Glassman, Kahn, in prep.) and two genera, *Aphandra* and *Asterogyne*, are new for Amazonia. The following estimate of the number of Amazonian palm species in each genus takes into account the currently valid species' names, the preliminary statement of Henderson's palm flora of the Amazon basin (in prep.), and the authors' point of view (Table 1).

The Amazonian palm flora is composed of about 180 species in 39 native genera, of which 33 occur in Peru, 28 in Brazil (Henderson and Balick 1987), with ca. 150 species in both cases, 24 genera with ca. 70 species in Ecuador (Balslev and Barfod 1987; Barfod 1991), and 18 genera with 67 species in French Guiana (de Granville 1990).

1.2 The Major Groups of the Amazonian Palms

Five of the six subfamilies defined by Dransfield and Uhl (1986) have Amazonian representatives.

Coryphoideae – Only four species in two of the 39 genera of Coryphoideae are found in Amazonia: three species of *Chelyocarpus* [*C. ulei* Dammer, *C. repens* Kahn & Mejia, *C. chuco* (Martius) H.E. Moore], and one of *Itaya* (*I. amicorum* H.E. Moore).

Table 1. The native palm genera with an estimate number of 180–190 Amazonian species

Coryphoideae

– Corypheae
  Thrinacinae – *Chelyocarpus* (3), *Itaya* (1)

Calamoideae

– Lepidocaryeae
    Mauritia (2), *Mauritiella* (1–3), *Lepidocaryum* (1–3)
– Calameae
    Raphia (1)

Ceroxyloideae

– Hyophorbeae
    Chamaedorea (5), *Wendlandiella* (3)

Arecoideae

– Iriarteeae
  Iriarteinae – *Dictyocaryum* (2), *Iriartella* (2), *Iriartea* (1), *Socratea* (2)
  Wettiniinae – *Catoblastus* (1), *Wettinia* (2)
– Areceae
  Manicariinae – *Manicaria* (1–3)
  Leopoldiniinae – *Leopoldinia* (3)
  Euterpeinae – *Euterpe* (3), *Prestoea* (4), *Oenocarpus* (9), *Jessenia* (1), *Hyospathe* (1)
– Cocoeae
  Butiinae – *Syagrus* (5)
  Attaleinae – *Attalea* (1), *Scheelea* (ca. 6), *Orbignya* (ca. 5), *Maximiliana* (1)
  Elaeidinae – *Barcella* (1), *Elaeis* (1)
  Bactridinae – *Acrocomia* (1), *Bactris* (50), *Aiphanes* (5), *Desmoncus* (ca. 6),
    Astrocaryum (21)
– Geonomeae
    Pholidostachys (1), *Geonoma* (30), *Asterogyne* (1)

Phytelephantoideae

  Ammandra (1), *Aphandra* (1), *Phytelephas* (1)

Calamoideae – Only four of the 22 genera of Calamoideae occur in Amazonia: *Mauritia*, *Mauritiella*, *Lepidocaryum* (Lepidocaryeae), with 3, 14, 9 still valid species, respectively (the former with probably two species, the other with only one or two species, after reassessment of these genera), and *Raphia* (Calameae) with just one species, *R. taedigera* (Martius) Martius, of the approximately 28 species recognized.

Ceroxyloideae – Two of the 11 genera are found in Amazonia: *Chamaedorea* and *Wendlandiella* (Hyophorbeae), the former with about five Amazonian species, and the latter with three species (*W. gracilis* Dammer, *W. polyclada* Burret, *W. simplicifrons* Burret), all in Peruvian Amazonia.

Arecoideae – Of the 122 genera of this subfamily, 28 occur in Amazonia. All six genera of the Iriarteeae are found in Amazonia, represented by the following species: Iriarteinae – *Dictyocaryum* [*D. lamarckianum* (Martius) H.A. Wendland, *D. ptariense* (Steyermarck) Moore & Steyermark], *Iriartella* [*I. setigera* (Martius) H.A. Wendland, *I. stenocarpa* Burret], *Iriartea* (*I. deltoidea* Ruiz & Pavon), and *Socratea* [*S. exorrhiza* (Martius) H.A. Wendland and *S. salazarii* H.E. Moore]; Wettiniinae – *Catoblastus* (*C. drudei* Cook & Doyle), *Wettinia* (*W. augusta* Poeppig & Endlicher, *W. maynensis* Spruce). Of the 86 genera of Areceae, only seven occur in Amazonia; these include: Manicariinae – *Manicaria* (*M. atricha* Burret, *M. martiana* Burret, *M. saccifera* Gaertner), there may be only one species with high variability (Uhl and Dransfield 1987); Leopoldiniinae – *Leopoldinia* (with probably only three species: *L. major* Wallace, *L. piassaba* Wallace, *L. pulchra* Martius); Euterpeinae – *Euterpe* (with at least three species, *E. oleracea* Martius, *E. precatoria* Martius, and *E. catinga* Wallace), *Prestoea* [with four species, *P. asplundii* H.E. Moore, *P. ensiformis* (Ruiz & Pavón) Martius, *P. megalochlamys* (Burret) H.E. Moore, *P. schultzeana* (Burret) H.E. Moore], *Oenocarpus* (with nine species, *O. bacaba* Martius, *O. balickii* Kahn, *O. circumtextus* Martius, *O. discolor* Barbosa Rodrigues, *O. distichus* Martius, *O. macrocalyx* Burret, *O. mapora* Karsten, *O. minor* Martius, *O. tarampabo* Martius), *Jessenia* [with one species and two subspecies, *J. bataua* (Martius) Burret subsp. *bataua* and *J. bataua* subsp. *oligocarpa* (Grisebach & H.A. Wendland) Balick], *Hyospathe* (with only one species, *H. elegans* Martius). Of the 22 genera of Cocoeae, 12 are found in Amazonia: Butiinae – *Syagrus* [*S. cocoides* Martius, *S. inajai* (Spruce) Beccari, *S. sancona* Karsten, *S. smithii* (H.E. Moore) Glassman, *S. stratincola* Wessels Boer]; Attaleinae – *Attalea* (probably with only one species, *A. tessmannii* Burret), *Scheelea* (probably no more than six species), *Orbignya* (probably no more than five species), *Maximiliana* [with only one species, *M. maripa* (Correa de Serra) Drude]; Elaeidinae – *Elaeis* [*E. oleifera* (Humboldt, Bonpland & Kunth) Cortes], *Barcella* [*B. odora* (Trail) Drude]; Bactridinae – *Acrocomia* (with one species, *A. lasiospatha* Martius), *Aiphanes* (about five species), *Bactris* (about 90 species have been described from Amazonia, probably less than 50 .species will be retained), *Desmoncus* (probably no more than six species), *Astrocaryum* (21 species). Of the six genera of Geonomeae, three are Amazonian palms: *Pholidostachys* [*P. synanthera* (Martius) H.E. Moore], *Asterogyne* (*A. guianensis* de Granville & Henderson), and *Geonoma* (about 30 species).

Phytelephantoideae – The genera *Phytelephas* (*P. macrocarpa* Ruiz & Pavon), *Ammandra* [*A. dasyneura* (Burret) Barfod] and *Aphandra* [*A. natalia* (Balslev & Henderson) Barfod] occur in Amazonia.

1.3 Palm Distribution in Amazonia

1.3.1 Spruce's Palm Regions

Spruce (1871, p. 69–71) was the first to discuss the palm distribution in the Amazon basin. He distinguished five palm regions: (1) The "Coast- or Submaritime Region, viz. the country adjacent to the mouth of the great river, both terra firma and islands, as far inland (or westward) as there are tidal creeks and the seabreezes have a manifest influence on the vegetation. This region ought to include the whole of the Guayana coast, the mouth of the Orinoco . . .". (2) The "Granite Region, . . . it belongs equally to the Rio Negro and to the Orinoco, extending down the former nearly to lat. 2°S., and down the latter to and beyond the cataracts (lat. 6°N.), the Casiquiari is its middle term, from which it stretches eastward through Guayana as far as to the falls of the rivers that run direct to the Atlantic, and westward nearly (or quite) to the foot of the Andes". (3) The "Diamond-Region – the elevated rocky region of Central Brazil, where the largest affluents of the Amazon take their rise . . .". (4) The "Amazon region – middle and upper – comprising the whole course of the main river and the country adjacent to its banks, from the foot of the Andes down to the commencement of the Pará archipelago, or westward limit of the Coast-Region". (5) The "Subandine region, comprising the eastern slopes of the Andes of Peru and Ecuador, up to 6000 feet, with a broad strip of the great plain at their base". Then the author commented upon palms of these regions, except for the "Diamond-Region" he had not visited.

In order to report and describe palm distribution, Amazonia will be divided into three parts from East to West, each including a northern and southern region. These are not "palm regions" and must not be understood as any strictly defined phytogeographic units. (1) The eastern region corresponds to Spruce's "Coast-Region" with a larger extension westwards. The northern part includes the Brazilian territory of Amapa, French Guiana, Surinam, and coastal parts of Guyana. The southern part, from the Atlantic coast to around Santarém, does not extend westwards beyond the Xingú River valley. (2) The western region corresponds to Spruce's "Subandine Region" but includes a broader strip of lowlands which extends to the Brazilian region of Tefé and covers Acre, Peruvian Madre de Dios and Bolivian Amazonia to the South, and the western part of his "Granite Region" to the North. As a matter of fact, most palm genera of the Andean piedmont and mountain slopes, which belong to a Subandean flora (*Aiphanes*, *Ammandra*, *Aphandra*, *Chamaedorea*, *Chelyocarpus*, *Dictyocaryum*, *Iriartea*, *Prestoea*, *Wettinia*), occur in the western lowlands of the Amazon basin. (3) The central region of the Amazon basin includes the Rio Negro valley, the Brazilian Territory of Roraima and Amazonian parts of Colombia, Venezuela, and Guyana to the North, and extends from the Xingú to the Madeiras River to the South. This area corresponds to Spruce's "Granite Region" northwards, to the central part of his "Amazon region", and to his "Diamond-Region" southwards.

1.3.2 Distribution of Genera in Amazonia

Of the 39 Amazonian palm genera, 34 (87.2%), 22 (56.4%), and 20 (51.3%) occur in the western, central, and eastern region, respectively. Seventeen genera are found throughout the Amazon basin (Table 2): *Astrocaryum*, *Bactris*, *Desmoncus*, *Elaeis*, *Euterpe*, *Geonoma*, *Hyospathe*, *Jessenia*, *Manicaria* (which was recently collected in Peruvian Amazonia by K. Mejia), *Mauritia*, *Mauritiella*, *Maximiliana*, *Oenocarpus*, *Orbignya*, *Scheelea*, *Socratea*, and *Syagrus*; 15 genera are located in western Amazonia: *Aiphanes*, *Ammandra*, *Aphandra*, *Attalea*, *Catoblastus*, *Chamaedorea*, *Chelyocarpus*, *Dictyocaryum*, *Iriartea*, *Itaya*, *Pholidostachys*, *Phytelephas*, *Prestoea*, *Wendlandiella*, and *Wettinia*; some reach central Amazonia: *Iriartella* and *Lepidocaryum*; two genera display limited distributions in northern part of central Amazonia: *Barcella* and *Leopoldinia*; *Acrocomia* occurs in central and eastern Amazonia: *Asterogyne* and *Raphia* are located only in the eastern part.

The distributions of some genera are particularly restricted. For example, *Asterogyne* was recently found in a very small area near the southern border of French Guiana (de Granville and Henderson 1988). *Chelyocarpus* and *Itaya* form infrequent populations, *Wendlandiella* is only known to grow in Peru, and *Ammandra* and *Aphandra* only in Ecuador and adjacent Colombia and Peru (Barfod 1991). Such distribution patterns have been discussed (Moore 1973; de Granville and Henderson 1988) with respect to the refugia theory (Haffer 1969; Prance 1973, 1982a,b).

1.3.3 Distribution of Species in Amazonia

In the eastern region, the characteristic native species are *Acrocomia lasiospatha*, *Asterogyne guianensis*, *Astrocaryum aculeatum* Meyer, *A. murumuru* Martius, *A. paramaca* Martius, *A. sciophilum* (Miquel) Pulle, *A. vulgare* Martius, *Bactris aubletiana* Trail, *B. cruegeriana* Grisebach & Wendland ex Grisebach, *B. oligocarpa* Barbosa Rodrigues, *B. oligoclada* Burret, *B. penicillata* Barbosa Rodrigues, *B. rhaphidacantha* Wessels Boer, *Euterpe oleracea*, *Geonoma euspatha* Burret, *G. oldemanii* de Granville, *G. pauciflora* Martius, *G. poiteauana* Kunth, *G. stricta* (Poiteau) Kunth, *G. triglochin* Burret, *Jessenia bataua* subsp. *oligocarpa*, *Manicaria saccifera*, *Orbignya sagotii* Trail ex Im Thurn, *Syagrus stratincola*; other species

Table 2. Distribution of palm genera in Amazonia

	North			South		
	W	C	E	W	C	E
Acrocomia	–	+	+	–	+	+
Aiphanes	+	–	–	+	–	–
Ammandra	+	–	–	+[a]	–	–
Aphandra	+	–	–	–	–	–
Asterogyne	–	–	+	–	–	+
Astrocaryum	+	+	+	+	+	+
Attalea	+	–	–	+	–	–
Bactris	+	+	+	+	+	+
Barcella	–	+	–	–	–	–
Catoblastus	–	–	–	+	–	–
Chamaedorea	+	–	–	+	–	–
Chelyocarpus	+	–	–	+	–	–
Desmoncus	+	+	+	+	+	+
Dictyocaryum	–	–	–	+	–	–
Elaeis	+	+	+	+	+	+
Euterpe	+	+	+	+	+	+
Geonoma	+	+	+	+	+	+
Hyospathe	+	+	+	+	+	+
Iriartea	+	–	–	+	–	–
Iriartella	+	+	–	+	+	–
Itaya	+	–	–	–	–	–
Jessenia	+	+	+	+	+	–
Leopoldinia	–	+	–	–	–	–
Lepidocaryum	+	+	–	+	–	–
Manicaria	+	+	+	–	–	–
Mauritia	+	+	+	+	+	+
Mauritiella	+	+	+	+	+	+
Maximiliana	+	+	+	+	+	+
Oenocarpus	+	+	+	+	+	+
Orbignya	+	+	+	+	+	+
Pholidostachys	+	–	–	+	–	–
Phytelephas	+	–	–	+	–	–
Prestoea	+	–	–	+	–	–
Raphia	–	–	–	–	–	+
Scheelea	+	+	–	+	+	–
Socratea	+	+	+	+	+	+
Syagrus	+	+	+	+	+	+
Wendlandiella	–	–	–	+	–	–
Wettinia	+	–	–	+	–	–
Total	31	22	18	31	18	17

W: west; C: central; E: east. [a] Probably in Acre, Brazil (Barfod 1991).

reach central Amazonia, such as *Astrocaryum gynacanthum* Martius, *A. rodriguesii* Trail, *Bactris elegans* Barbosa Rodrigues & Trail ex Barbosa Rodrigues, *B. gastoniana* Barbosa Rodrigues, *B. trailiana* Barbosa Rodrigues, *Desmoncus mitis* Martius, *Geonoma baculifera* (Poiteau)

Kunth, *Oenocarpus bacaba*, *Orbignya spectabilis* (Martius) Burret, *Syagrus inajai*. In the southern part, there are *Raphia taedigera*, *Oenocarpus distichus* and *Orbignya phalerata* Martius; this latter species is widely distributed from the eastern to the southwestern region of the Amazon basin.

In the north of the central region, along the Rio Negro valley, the most characteristic palm species are *Astrocaryum acaule* Martius, *A. sociale* Barbosa Rodrigues, *Bactris bidentula* Spruce, *Geonoma arundinacea* Martius, *G. multiflora* Martius, *Leopoldinia* spp., *Lepidocaryum gracile* Martius, *L. tenue* Martius, *Manicaria martiana*, and *Oenocarpus minor*. Several species frequent in central Amazonia also occur in the western region, such as *A. javarense* (Trail) Trail ex Drude, *Bactris chaetochlamys* Burret, *B. mitis* Martius, *Euterpe catinga*, *G. densiflora* Spruce, *G. laxiflora* Martius, *G. spixiana* Martius, *Iriartella setigera*, *Jessenia bataua* subsp. *bataua*, *Mauritia carana* Wallace, *Oenocarpus mapora*, and *Scheelea insignis* (Martius) Karsten. *Acrocomia lasiospatha* and *Astrocaryum aculeatum*, both eastern species, are likely to have been introduced in central Amazonia. No peculiar palm species are characteristic of the southern part of the central region. *Acrocomia totai* Martius, *Astrocaryum chonta* Martius, *Scheelea princeps* (Martius) Karsten, and *Syagrus sancona* are particularly abundant in the drier forests and savannahs. *Copernicia alba* Morong ex Morong & Britton found in association with the former species, and *Trithrinax schizophylla* Drude mark the southern limit of the Amazonian palm flora in Matto Grosso, Brazil, and extend westwards to Bolivia.

In the western region, which is under the influence of both Subandean and Amazonian flora, the most characteristic species are *Aiphanes deltoidea* Burret, *A. tessmannii* Burret, *A. ulei* (Dammer) Burret, *Ammandra dasyneura*, *Aphandra natalia*, *Astrocaryum chambira* Burret, *A. huicungo* Burret, *A. urostachys* Burret, *Bactris acanthospatha* (Trail) Trail ex Drude, *B. bifida* Martius, *Chamaedorea pauciflora* Martius, *C. lanceolata* (Ruiz & Pavon) Kunth, *C. pinnatifrons* (Jacquin) Oersted, *Chelyocarpus chuco*, *C. repens*, *C. ulei*, *Desmoncus leptospadix* Martius, *D. longifolius* Martius, *Dictyocaryum lamarckianum* (only found at the western limit on eastern mountains of the Andes), *D. ptariense*, *Geonoma brongniartii* Martius, *G. camana* Trail, *G. helminthoclada* Burret, *G. interrupta* (Ruiz & Pavon) Martius, *G. juruana* Dammer, *G. jussieuana* Martius, *G. macrostachys* Martius, *G. peoppigiana* Martius, *Iriartea deltoidea*, *Iriartella stenocarpa*, *Itaya amicorum*, *Lepidocaryum tessmannii* Burret (probably only one species with *L. gracile* and *L. tenue*), *Pholidostachys synanthera*, *Scheelea bassleriana* Burret, *S. brachyclada* Burret, *S. cephalotes* (Poeppig ex Martius) Karsten, *S. stenorhyncha* Burret, *S. tessmannii* Burret, *Socratea salazarii*, *Wendlandiella gracilis*, *W. polyclada*, *W. simplicifrons*, *Wettinia augusta*, and *W. maynensis*.

Several palm species are widely distributed throughout the basin, from the Atlantic coast to the Andean piedmont, from North to South. These include: *Astrocaryum jauari* Martius, *Bactris acanthocarpoides* Barbosa Rodrigues, *B. concinna*, *B. constanciae* Barbosa Rodrigues, *B. humilis* (Wallace) Burret, *B. maraja* Martius, *B. monticola* Barbosa Rodrigues, *B. pectinata* Martius, *B. simplicifrons* Martius, *Desmoncus macroacanthos* Martius, *D. orthacanthos* Martius, *D. polyacanthos* Martius, *Elaeis oleifera*, *Euterpe precatoria*, *Geonoma acaulis* Martius (less fequent in the East), *G. bartlettii* Burret, *G. deversa* (Poiteau) Kunth, *G. leptospadix* Trail, *G. maxima* (Poiteau) Kunth, *G. piscicauda* Dammer, *G. pycnostachys* Martius, *Hyospathe elegans*, *Mauritia flexuosa* L.f., *Maximiliana maripa*, *Orbignya polysticha* Burret, *Socratea exorrhiza*. *Mauritiella aculeata* (Humboldt, Bonpland & Kunth) Burret is absent from French Guiana but present in the other regions.

1.4 Conclusion

Palms are well represented in the Amazon basin with 20% of the genera. Palm distribution is not homogeneous throughout the basin. They are particularly diversified in the western region where Amazonian and Subandean floras are superimposed. Sixteen genera (41.0%) are represented by only one species. Nineteen genera (48.7%) have probably less than six species. Only four of the thirty-nine Amazonian genera, *Astrocaryum*, *Bactris*, *Geonoma*, and *Oenocarpus*, are more diversified, and total together more than 50% of the total number of species. Extensive areas of the basin have not yet been collected, however, and the jungle may have some good surprises in reserve for each new botanical expedition.

2 Life Forms of Amazonian Palms

2.1 Growth Models of Amazonian Palms

Four of the growth models defined by Hallé et al. (1978) are developed by palms. These include: Holttum's model, i.e., a hapaxanthic (inflorescence from the apical meristem), unbranched axis; Corner's model, i.e., a pleonanthic (inflorescences from lateral buds), unbranched axis; Tomlinson's model, i.e., hapaxanthic or pleonanthic palms with basal branching; Schoute's model, i.e., pleonanthic axes branched by equal dichotomy. Some species combine elements of two architectural models (Tomlinson 1990). This is the case of the Amazonian *Dictyocaryum ptariense* and *Socratea salazarii*, which are single-trunked palms (Corner's model) in spite of producing stolons from their base (Tomlinson's model). The other Amazonian species develop either Corner's or Tomlinson's model, and all of them are pleonanthic, except for *Raphia taedigera* which is a hapaxanthic species; this latter form is frequent in other continents (Dransfield 1978). From these two strategies of growth, Amazonian palms have developed many different life forms.

2.2 What Kind of Plants Are the Amazonian Palms?

Large, arborescent, erect-trunked palms bestow on the family its symbol, so often represented by a coconut tree on a splendid beach. However, this is but one of the various life forms developed by the family. Palm life forms will be defined from the presence or absence (acaulescent) of a stem. Within the stem category, further distinction is made according to stem type (aerial: erect, prostrate or climbing, or subterranean), basal branching potential (single or multistemmed species), and adult height of the palm (large: over 12 m, medium: 6 to 12 m, and small: less than 6 m in height). Comments will be made on the leaf morphology, the stem which is unarmed or strongly armed, smooth or covered by sheaths of dead leaves, cylindrical or swollen, and upon stilt roots and pneumatophores. The major features of palm anatomy and "structural biology" are summed up by Tomlinson (1990).

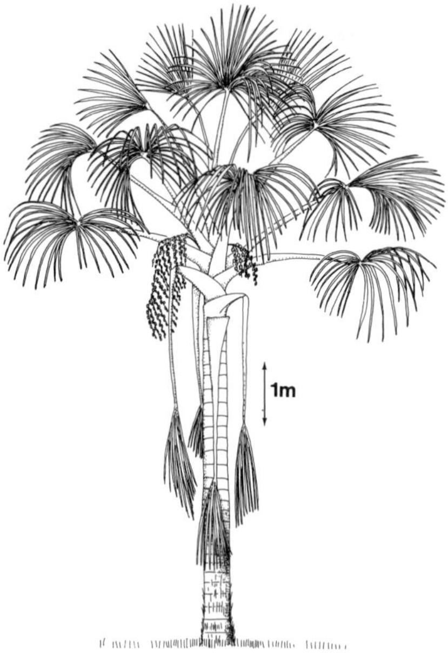

Fig. 2. *Mauritia flexuosa*

2.2.1 Description of Palm Life Forms

2.2.1.1 Large, Arborescent Palms with Erect Trunk

The leaves of the tallest palms do not reach the crown of the tallest trees, and the image of palms emerging from forest canopy is somewhat misleading, as clearly noted by Spruce (1871, p. 82–83): "A traveller approaching by sea the sities of Panamá, Guayaquil, and many others within the tropics, will see groves of Coco-palms towering far above the bushy spreading Mangos and Ingas that nestle at their base; but the latter are by no means forest-trees, nor is the Coco a forest-palm. Let him, however, leave the coast and penetrate the virgin forest beyond, and he will see that the loftiest palms do not usually exceed the exogenous trees of average height, and that, except on the river-banks, they are often quite hidden from view until closely approached. From some of the naked-topped granite hills of the Rio Negro and Orinoco, and from the ascent of the eastern side of the Andes, I have looked over perfect oceans of forest, and am able to assert that very rarely do palms domineer over all other trees – so rarely, indeed, that I believe I have only noted it twice, and that on a very limited area, during the whole course of my travels."

The few species we found with crowns among the foliages of the highest trees, at about 30 m in height, were *Astrocaryum chambira*, *Iriartea deltoidea*, and *Scheelea bassleriana* in a Peruvian terra firme forest, and

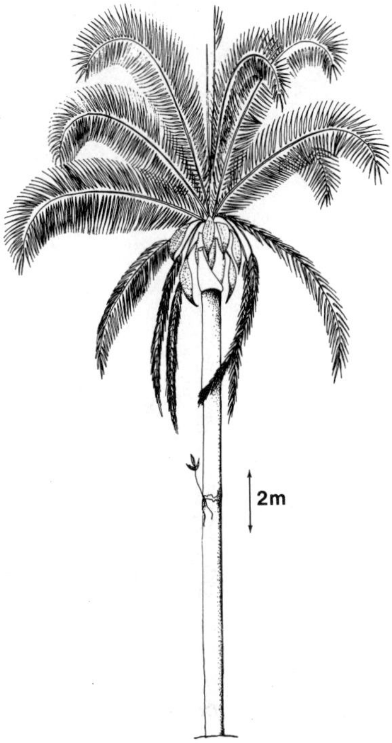

Fig. 3. *Scheelea bassleriana*

Mauritia flexuosa, which is the major component of swamp forest canopies through the basin. All the former species are single-stemmed palms. The arborescent, multistemmed species usually develop smaller leaves and stems.

Tall, Single-Stemmed Palms. The species with largest trunk are *Mauritia flexuosa* (Fig. 2) and *Scheelea* spp. (Fig. 3), with DBH up to 50 cm, *Iriartea deltoidea*, *Astrocaryum aculeatum*, *A. chambira* (Fig. 4), *Maximiliana maripa* (Fig. 35, Chap. 3), *Orbignya phalerata*, and *Jessenia bataua* (Fig. 40, Chap. 3), these with DBH up to 30 cm. The leaves of *Maximiliana maripa* present a high number of pinnae, 140 to 260 pairs, and reach up to 11 m in length (Glassman, in prep.). The leaves of *Jessenia bataua* (including rachis and petiole) attain a maximum length of 9 m, with up to 106 pairs of pinnae, up to 1.56 m long and 13 cm wide (Table 3). The leaves of *Astrocaryum chambira* are shorter, maximum 6.7 m long, with up to 154 pairs of pinnae, which can reach 1.4 m in length and 4.7 cm in width (Table 4). Tall palms which can develop a slender trunk include the following species: *Euterpe precatoria* (Fig. 5), *Dictyocaryum ptariense*, *Socratea exorrhiza* (Fig. 6), and *Syagrus sancona*. The leaf length of *Euterpe precatoria* is less than 4 m with up to 103 pairs of pinnae. There is substantial variability between populations, however, as is shown by comparative data from central and

Fig. 4. *Astrocaryum chambira*

Fig. 5. *Euterpe precatoria*

western Amazonia; leaf size of *E. precatoria* in a Peruvian population is clearly greater than in a Brazilian population (Table 5). While leaf length of *Dictyocaryum ptariense* and *Socratea exorrhiza* in Peruvian populations is less than 3.5 m (Table 6), both species attain a height of over 20 m.

Tall, Multistemmed Palms. This life form is developed only by a few Amazonian species, *Astrocaryum jauari*, *A. vulgare*, *Euterpe oleracea* (Fig. 7), *Mauritiella aculeata* (Fig. 8). These species present a high variability in size, and in some cases, they could be included in the medium-sized category.

2.2.1.2 Medium-Sized Palms with Erect Stem

Two forms are recognized, each with single and multistemmed species. One is characterized by a slender stem, 5–10 cm in diameter, and rather short leaves, no more than 4 m in length, and includes the following species: *Socratea salazarii*, *Oenocarpus balickii*, *Syagrus inajai* (Fig. 59, Chap. 5), and *Wettinia maynensis*, all single-stemmed; *Oenocarpus mapora*, *O. minor*, *Phytelephas macrocarpa*, *Syagrus stratincola*, *Wettinia augusta* which are multistemmed species (Table 7). The other which is characterized by a

Table 3. *Jessenia bataua* subsp. *bataua*, a large, single-stemmed palm (data from Peruvian Amazonia)

L	No. of pinnae per side	Rachis length (cm)	Median pinna Length (cm)	Median pinna Width (cm)	DBH (cm)
6	89.5 ± 3.4	508.8 ± 13.5	120.0 ± 4.9	10.4 ± 0.5	21.0
6	92.5 ± 10.1	580.7 ± 38.4	127.2 ± 8.1	10.9 ± 0.9	18.8
8	92.4 ± 2.1	546.8 ± 8.4	139.1 ± 8.1	13.3 ± 0.8	23.6
6	104.2 ± 5.3	767.8 ± 20.7	130.4 ± 10.8	11.0 ± 0.6	27.1
3	94.3 ± 3.8	450.0 ± 6.1	85.7 ± 2.6	7.0 ± 0.4	19.6
5	102.4 ± 1.9	729.4 ± 31.0	156.0 ± 1.3	10.6 ± 0.7	22.6
6	100.0 ± 3.7	747.2 ± 24.1	138.7 ± 5.6	13.1 ± 0.6	20.7
5	100.8 ± 2.4	687.6 ± 13.4	125.2 ± 5.4	10.1 ± 0.5	19.7
5	92.8 ± 1.2	658.2 ± 34.8	119.6 ± 4.8	9.1 ± 0.7	19.4
7	98.9 ± 2.6	659.9 ± 10.0	117.6 ± 4.7	12.0 ± 1.1	20.7
6	87.8 ± 1.7	485.5 ± 14.3	122.7 ± 4.6	10.5 ± 1.6	25.5
6	95.5 ± 1.7	719.0 ± 25.7	138.8 ± 3.2	12.0 ± 0.1	21.0
5	103.4 ± 3.6	587.6 ± 13.1	120.0 ± 2.5	10.3 ± 0.8	20.7
3	97.7 ± 3.4	773.7 ± 19.4	100.0 ± 1.0	9.7 ± 0.4	18.1
5	102.8 ± 7.4	714.2 ± 44.8	137.6 ± 12.5	11.9 ± 1.0	21.3
3	90.0 ± 9.2	433.0 ± 27.1	108.3 ± 4.0	9.4 ± 0.4	18.1
9	103.3 ± 2.9	631.3 ± 19.5	137.4 ± 5.7	12.3 ± 1.0	21.6
7	102.9 ± 3.6	676.3 ± 27.4	146.3 ± 7.5	9.7 ± 0.9	24.8
9	100.6 ± 3.7	713.1 ± 41.4	145.2 ± 4.3	12.1 ± 0.8	25.5
7	92.9 ± 4.7	709.1 ± 21.9	148.0 ± 7.9	9.9 ± 1.0	24.2
6	97.5 ± 5.1	701.7 ± 43.5	143.0 ± 6.9	11.3 ± 0.6	19.4
11	100.7 ± 3.7	658.4 ± 16.9	123.2 ± 10.5	9.5 ± 0.7	20.4
10	97.2 ± 1.5	715.9 ± 33.2	134.4 ± 6.9	11.0 ± 0.9	18.8
11	93.7 ± 5.0	529.2 ± 36.9	108.1 ± 4.8	9.7 ± 0.8	19.7
8	106.3 ± 4.7	871.5 ± 61.4	119.1 ± 5.6	11.3 ± 1.1	20.7
9	103.0 ± 2.9	730.6 ± 20.8	142.9 ± 4.4	11.4 ± 0.9	21.0
10	92.2 ± 2.7	588.7 ± 19.0	135.7 ± 4.5	12.1 ± 0.6	25.8
11	100.5 ± 3.6	636.8 ± 16.0	133.6 ± 4.3	9.6 ± 0.7	22.0
11	98.7 ± 2.6	657.6 ± 11.1	133.5 ± 5.3	10.9 ± 1.0	24.8
9	103.1 ± 3.2	707.9 ± 17.1	123.6 ± 4.8	10.3 ± 1.2	16.6
M	97.9 ± 5.1	652.6 ± 102.0	128.7 ± 15.3	10.7 ± 1.3	21.4 ± 2.6

L: Number of leaves measured per palm from which means (± standard deviation) are calculated; M: mean calculated from the 30 palms; trunk length: 8.1–25.8 m.

Table 4. *Astrocaryum chambira*, a large, single-stemmed palm (data from Peruvian Amazonia)

L	No. of pinnae per side	Rachis length (cm)	Sheath and petiole length (cm)	Median pinna Length (cm)	Median pinna Width (cm)
14[a]	154.1 ± 6.9	361.8 ± 21.7	292.4 ± 13.9	126.8 ± 6.3	4.5 ± 0.7
14[b]	144.8 ± 13.4	356.4 ± 17.0	320.4 ± 27.1	144.3 ± 16.1	4.7 ± 0.9
11[c]	129.9 ± 5.6	350.1 ± 17.5	260.9 ± 14.7	126.0 ± 5.0	4.7 ± 0.4

L: Number of leaves measured per palm from which means (± standard deviation) are calculated. Three palms sampled in a deforested area. [a] Height 5 m, DBH 33 cm. [b] Height 7 m, DBH 40 cm. [c] Height 7 m, DBH 38 cm.

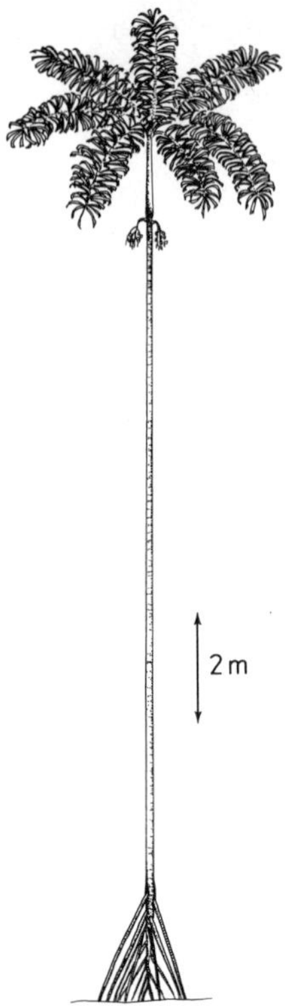

Fig. 6. *Socratea exorrhiza*

Fig. 7. *Euterpe oleracea*

Fig. 8. *Mauritiella aculeata*

Table 5. Tall, slender, single-stemmed palms

Euterpe precatoria

L	No. of pinnae per side	Rachis length (cm)	Median pinna Length (cm)	Median pinna Width (cm)	Trunk Length (m)	Trunk DBH (cm)
A) In central Amazonia (Brazil)						
6	69.7 ± 2.9	219.2 ± 5.3	84.6 ± 2.9	3.0 ± 0.1	15.2	10.4
7	68.3 ± 2.4	217.0 ± 3.5	64.4 ± 1.5	3.1 ± 0.1	10.6	9.8
7	78.0 ± 1.0	277.6 ± 6.3	100.4 ± 1.3	3.1 ± 0.1	15.5	15.0
6	60.6 ± 1.3	164.4 ± 3.8	66.0 ± 2.7	2.1 ± 0.1	18.1	11.0
7	64.3 ± 1.2	227.3 ± 10.5	68.7 ± 1.7	2.9 ± 0.1	10.4	10.1
7	66.0 ± 1.0	211.7 ± 2.7	76.1 ± 1.6	2.6 ± 0.1	15.3	9.9
7	69.9 ± 1.9	207.0 ± 3.2	80.4 ± 2.7	2.6 ± 0.1	19.9	13.7
Mean[a]	68.1 ± 5.4	217.5 ± 32.2	77.3 ± 12.1	2.8 ± 0.4	–	–
B) In western Amazonia (Peru)						
16	97.4 ± 4.0	326.7 ± 6.2	100.9 ± 4.6	2.7 ± 0.3	15.7	16.9
11	85.1 ± 2.2	179.0 ± 5.2	66.0 ± 5.9	2.1 ± 0.3	24.8	16.6
10	103.0 ± 4.1	324.3 ± 5.6	108.9 ± 3.2	3.0 ± 0.1	12.5	18.0
10	99.0 ± 4.0	334.6 ± 11.1	109.6 ± 3.5	3.2 ± 0.2	12.3	16.5
Mean[b]	96.0 ± 7.4	293.3 ± 64.3	96.2 ± 17.8	2.7 ± 0.5	–	–

L: Number of leaves measured per palm.
[a] Mean (± standard deviation) of the 47 leaves; 7 palms sampled: 10 to 16 contemporaneous leaves per palm; sheath length: 95 to 126 cm; petiole length: 41 to 46 cm (Kahn and Castro, unpubl. data).
[b] Mean of the 47 leaves; 4 palms sampled: 12 to 17 contemporaneous leaves per palm; sheath length: 80 to 167 cm; petiole length: 28 to 59 cm.

Table 6. Tall, slender, single-stemmed palms (data from Peruvian Amazonia)

L	No. of pinnae per side	Rachis length (cm)	Petiole length (cm)	Sheath length (cm)	Median pinna Length (cm)	Median pinna Width (cm)	DBH (cm)
Dictyocaryum ptariense							
2	39–41[a]	304–316	45	107–111	100.2[b] ± 10.2		
3	30–33	213–240	38–45	67–77	101.1 ± 5.9		
2	32–35	245–262	39–45	85–87	112.9 ± 6.7		
2	33–35	251–254	49–55	91–92	111.5 ± 10.8		
2	36–37	185–190	30–38	72–74	92.0 ± 4.5		
Mean		244.5 ± 41.0[c]	42.7 ± 6.6[c]	85.6 ± 15.0[c]	102.9 ± 10.6[d]		
Socratea exorrhiza							
Mean[e]	17.7[f] ± 1.7	215.4 ± 42.9			77.9[g] ± 12.4	22.2 ± 5.0	12.6 ± 1

L: Number of leaves measured per palm; 5 palms sampled: trunk length 14.1–24.0 m, DBH 18.5–22.9 cm.
[a] Each pinna is longitudinally divided in 2–8 segments.
[b] 3 segments from 3 median pinnae are measured per leaf.
[c] n = 11.
[d] n = 33.
[e] Mean (± standard deviation) calculated from 74 leaves measured from 16 palms sampled.
[f] Each pinna is longitudinally divided in 2–9 segments.
[g] 1 segment from 1 median pinna is measured per leaf.

Table 7. Medium-sized palms with slender stem (DBH < 12 cm) in Peruvian Amazonia

	No. of pinnae per side	Rachis length (cm)	Petiole length (cm)	Median pinna	
				Length (cm)	Width (cm)
Socratea salazarii[a]					
P/L					
1/1	16	273		86	22
1/1	15	195		79	15
1/1	16	172		–	–
1/8	14.0 ± 0	146.6 ± 5.9		65.4 ± 2.5	12.3 ± 0.8
16/109	15.8 ± 1.4	166.1 ± 21.7		73.8 ± 7.9	13.4 ± 1.8
Oenocarpus balickii[b]					
L					
3	89– 91	256–310		51–64	3.5–4.5
3	118–123	372–380		51–60	2.2–3.5
6	112–120	306–333		55–65	2.9–3.3
6	113–124	328–402		63–70	3.8–4.8
3	109–111	382–405		78–84	2.9–3.7
3	127–131	334–380		59–73	4.0–5.4
3	105–112	312–383		65–72	3.9–4.4
M[c]	112.8 ± 13.6	348.0 ± 39		65.0 ± 9.2	3.7 ± 0.7
Phytelephas macrocarpa					
L					
8	81.6 ± 5.4	291.7 ± 29.1	88.1 ± 14.1	57.3 ± 5.8	4.4 ± 0.3
2	72.5 ± 2.1	264.0 ± 22.6	109.3 ± 36.4	67.5 ± 0.5	4.5 ± 0.1
Wettinia augusta[d]					
P/L					
1/3	20–22	172–181		76–77	9–10
1/3	20–21	205–212		73–80	11–12
1/5	22–24	228–243		78–83	12–13
1/5	24–26	318–340		99–111	11–13
16/92	21.9 ± 2.0	231.3 ± 32.4		89.7 ± 9.9	11.8 ± 1.5

P: Number of palms sampled and L: number of leaves measured per palm(s).
[a] Stem length 4–8 m, DBH 7–12 cm; sheath length 82–122 cm, petiole length 30–57 cm.
[b] Stem length 8.0–11.2 m; DBH 7.5–8.0 cm; sheath length 40–60 cm, petiole length 0–15 cm.
[c] M: mean (± standard deviation) calculated from the 27 leaves measured.
[d] Stem length 8–12 m; DBH 7–9.2 cm; sheath length 68–119 cm, petiole length 22–39 cm.

larger diameter, 15–20 cm, and large leaves, 6–7 m in length, is developed by *Astrocaryum macrocalyx* Burret, *A. javarense* and *A. sciophilum* (Fig. 9), *Leopoldinia piassaba*, *Manicaria martiana* and *M. saccifera* (Fig. 10), all single-stemmed palms, and *Astrocaryum carnosum*, *A. murumuru* and *A. urostachys* all multistemmed species (Table 8A–D).

Table 8. Medium-sized palms with DBH > 12 cm and large leaves. A–D: erect trunk; E: prostrate trunk

P/L	No. of pinnae per side	Rachis length (cm)	Total length (cm)	Median pinna	
				Length (cm)	Width (cm)
A) *Astrocaryum macrocalyx*[a]					
8/24	100.0 ± 10.5	485.5 ± 62.7	691.1 ± 93.9	95.6 ± 10.9	4.3 ± 0.3
5/15	93.9 ± 8.6	482.3 ± 42.9	649.0 ± 37.9	104.5 ± 7.8	5.4 ± 0.6
B) *Astrocaryum javarense*[a]					
3/4	82–104	450–535	–	110–124	6.8–8.0
C) *Astrocaryum sciophilum*[b]					
24/24	73 ± 4.5		630 ± 35		
D) *Astrocaryum carnosum*[a]					
21/63	103.1 ± 11.1	478.5 ± 56.4	668.7 ± 66.4	100.9 ± 11.0	5.7 ± 0.8
E) *Elaeis oleifera*[a]					
30/30	82.0 ± 6.3	504.3 ± 42.6	765.2 ± 50.6	109.2 ± 12.8	5.8 ± 0.6

P: Number of palms sampled and L: total number of leaves measured in each population from which means (± standard deviation) are calculated.
[a] From Peruvian Amazonia.
[b] From French Guiana (Sist 1989b).

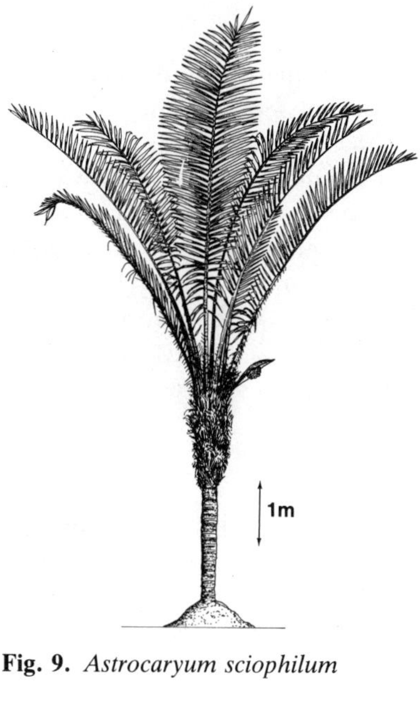

Fig. 9. *Astrocaryum sciophilum*

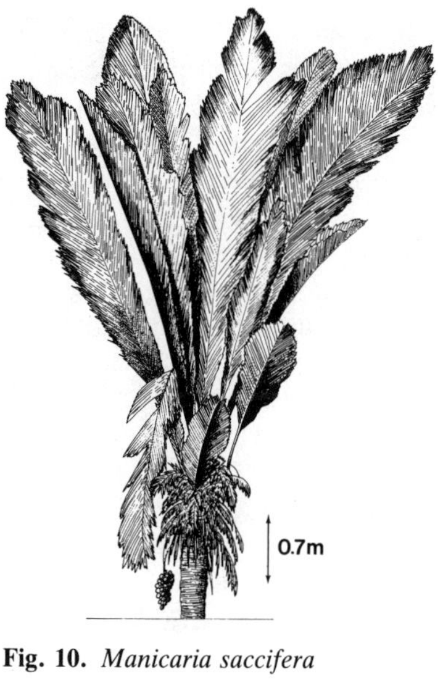

Fig. 10. *Manicaria saccifera*

Fig. 11. *Geonoma maxima*, 3 m in height

2.2.1.3 Small Palms with Erect Stem

The diameter of the stem is smaller than 6 cm, generally 1 to 3 cm. Leaf length does not exceed 2.5 m, and is generally between 1 and 2 m. Palm height is 0.7 to 6 m, most often less than 4 m. Most of them are multi-stemmed species (see Sect. 2.2.2). They are principally represented by the genera *Aiphanes*, *Asterogyne*, *Bactris*, *Chamaedorea*, *Geonoma*, *Hyospathe*, and *Wendlandiella* (Figs. 11, 12, 13). The smallest palms known in Amazonia belong to this category: *Bactris aubletiana*, *B. simplicifrons* (Fig. 14), *Geonoma stricta* (Fig. 15), and *Wendlandiella* spp., with a very slender stem, about 0.6 cm in diameter.

2.2.1.4 Prostrate Palms

There are a few species which develop a creeping stem. This runs along the ground and is erect only at the crown level. *Elaeis oleifera* (Fig. 16) is a large palm of small height because of its massive trunk which creeps up to 5 m, drawing a circle (Table 8E). *Chelyocarpus repens* is a small species with a short creeping stem up to 1 m long and 6–8 cm in diameter. Both species are

Fig. 12. *Geonoma leptospadix*, 2 m in height

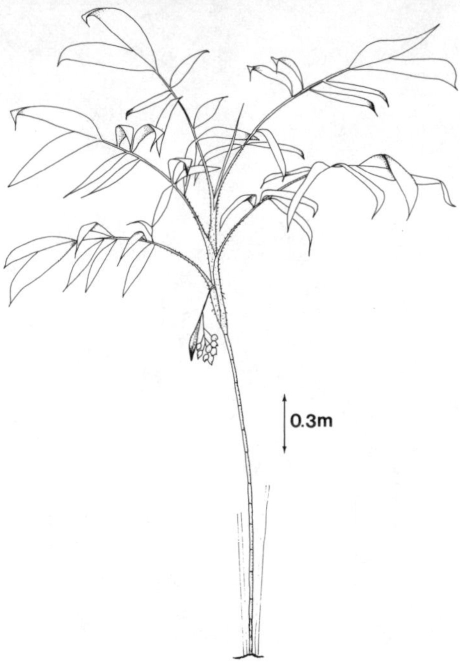

Fig. 13. *Bactris monticola*

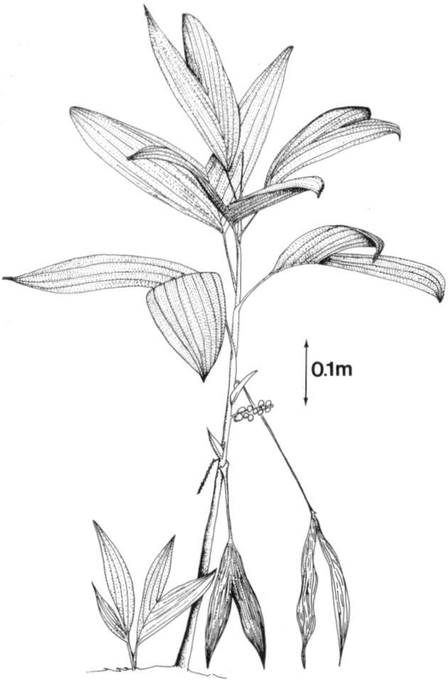

Fig. 14. *Bactris simplicifrons*

Fig. 15. *Geonoma stricta*, 0.7 m in height

Fig. 16. *Elaeis oleifera*, a prostrate palm

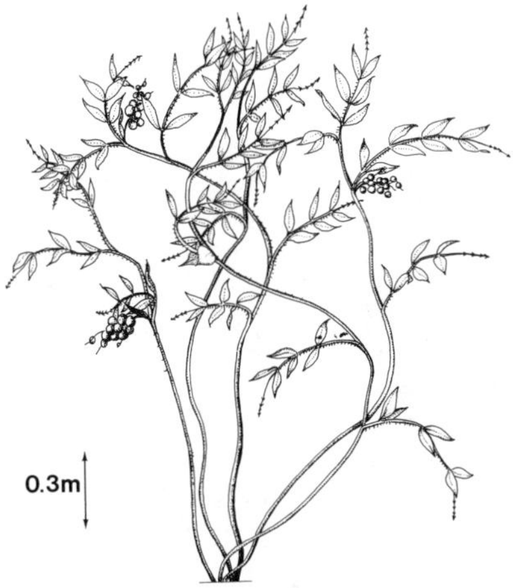

Fig. 17. *Desmoncus* sp., a climbing palm

single-stemmed palms. *Phytelephas macrocarpa* and *Ammandra dasyneura*, medium-sized and small palms respectively, can develop creeping stems. Both are multistemmed.

2.2.1.5 Climbing Palms

Desmoncus is the only genus which develops climbing palms in Amazonia. In comparison with Asian and African rattans, it is a poor example of lianescence (Fig. 17). *Desmoncus* spp. were never found over 15 m in height in Amazonia.

2.2.1.6 Subterranean-Stemmed Palms with Large Leaves

Palms with subterranean stems cannot be considered as acaulescent (Holttum 1955). This life form is characterized by a subterranean stem

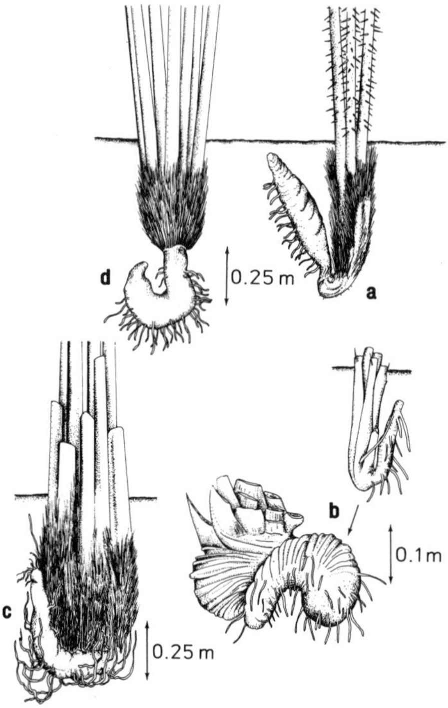

Fig. 18a–d. Subterranean-stemmed palms. **a** *Astrocaryum sociale*. **b** *A. paramaca*, juvenile and adult plants. **c** *Scheelea insignis*. **d** *Orbignya polysticha*

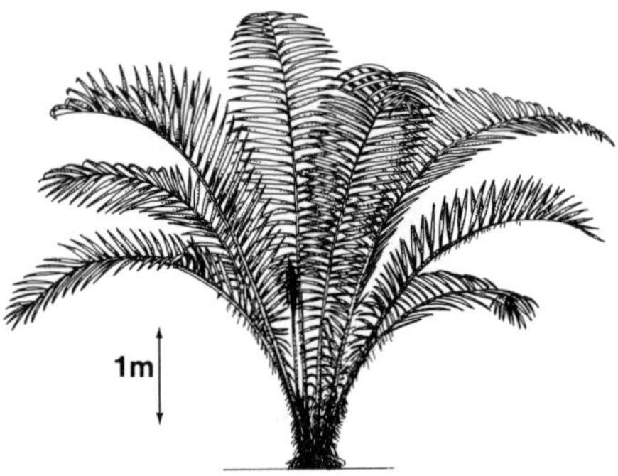

Fig. 19. *Astrocaryum paramaca*

which can reach 0.5 m in length and 15 cm in diameter (Fig. 18), and by a funnel-like crown of large leaves, up to 5 m long in *Astrocaryum sociale*, *A. paramaca* (Fig. 19), *Orbignya polysticha* (Fig. 20), *Orbignya sagotii*, *O. spectabilis* (Table 9), and up to 7 m long in *Astrocaryum acaule* and *Scheelea insignis*. The subterranean stem results from a positive geotropism (Corner 1966) and reaches a depth of 0.7–1 m. These subterranean-

Fig. 20. *Orbignya polysticha*, with 5-m-long leaves

stemmed, large-leafed species generally form dense populations and give the forest understory a particular physiognomy.

2.2.1.7 Subacaulescent Palms with Small Leaves

Several species with small leaves, no more than 2 m in length, produce no conspicuous stem. They include *Aiphanes ulei* (Fig. 21), *Bactris oligocarpa*, *G. acaulis*, *G. macrostachys* and *G. poiteauana*.

2.2.2 Basal Branching

Basal branching patterns produce either clumps of a few to many contemporaneous juvenile and adult axes or a single adult axis which will initiate another before degenerating. The terms "single-stemmed" and "multi-stemmed" species are employed here and not "solitary" and "clustered" or "clumped" species. A solitary axis may belong either to a potentially

Table 9. Leaf size of subterranean-stemmed palms

L	No. of pinnae per side	Rachis length (cm)	Total length (cm)	Petiole length (cm)	Median pinna	
					Length (cm)	Width (cm)
A) *Astrocaryum sociale* from central Amazonia, near Manaus						
7	51.7 ± 2.5	315.7 ± 10.4		130.4 ± 9.6	84.1 ± 3.8	2.9 ± 0.3
8	62.1 ± 7.0	300.9 ± 27.5		109.6 ± 12.4	82.4 ± 3.6	3.0 ± 0.2
7	69.9 ± 3.8	331.4 ± 26.0		134.4 ± 16.0	88.3 ± 8.2	3.1 ± 0.3
7	62.3 ± 2.6	334.6 ± 21.1		119.6 ± 13.5	76.6 ± 4.2	3.0 ± 0.3
8	60.1 ± 6.1	318.0 ± 24.7		144.9 ± 11.3	81.6 ± 6.8	3.7 ± 0.3
M[a]	61.2 ± 7.4	319.5 ± 24.9		127.8 ± 17.5	82.6 ± 6.5	3.2 ± 0.4
B) *Astrocaryum paramaca* from French Guiana (Sist 1989a)						
n=23 palms	82 ± 4.9		518 ± 31			
C) *Orbignya polysticha* from Peruvian Amazonia						
5	96.0 ± 2.3	432.2 ± 12.1		68.4 ± 4.0	74.0 ± 4.5	3.6 ± 0.2
6	96.3 ± 2.4	463.8 ± 20.8		118.0 ± 19.2	87.0 ± 7.3	3.5 ± 0.3
4	107.0 ± 1.4	457.9 ± 21.3		152.8 ± 21.4	82.8 ± 5.4	3.3 ± 0.2
4	95.0 ± 1.4	419.6 ± 7.1		125.8 ± 12.5	83.8 ± 3.1	4.0 ± 0.4
4	107.8 ± 4.0	501.5 ± 23.2		37.8 ± 14.7	78.6 ± 5.3	4.7 ± 0.1
6	91.2 ± 4.9	433.8 ± 40.3		90.9 ± 5.5	79.2 ± 5.5	3.4 ± 0.2
6	107.0 ± 1.5	447.1 ± 34.9		85.2 ± 19.0	77.1 ± 7.8	3.6 ± 0.2
6	94.2 ± 2.1	400.3 ± 9.3		88.6 ± 16.4	77.4 ± 5.1	3.3 ± 0.2
6	94.6 ± 2.1	389.5 ± 16.8		51.8 ± 9.2	69.5 ± 3.6	3.2 ± 0.2
5	107.2 ± 1.5	451.5 ± 27.5		91.9 ± 17.9	72.7 ± 4.6	3.6 ± 0.0
M[b]	99.1 ± 6.8	438.3 ± 37.5		90.4 ± 34.6	78.0 ± 7.2	3.6 ± 0.6

L: Number of leaves measured per palm.
[a] Mean (± standard deviation) calculated from the 37 leaves, 5 palms sampled; 8 to 9 contemporaneous leaves per crown.
[b] Mean calculated from the 52 leaves, 10 palms sampled; 5–7 contemporaneous leaves per crown.

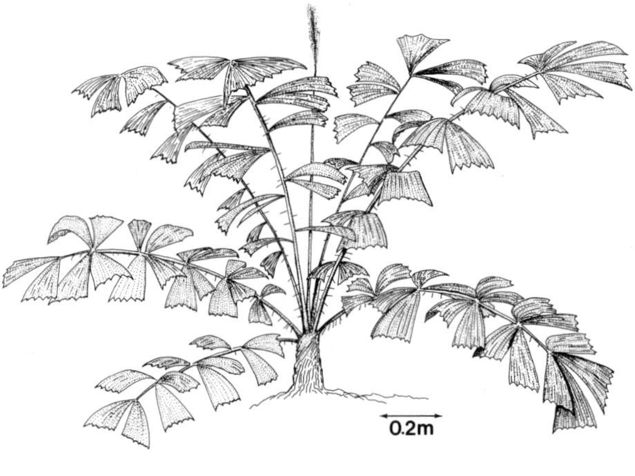

Fig. 21. *Aiphanes ulei*

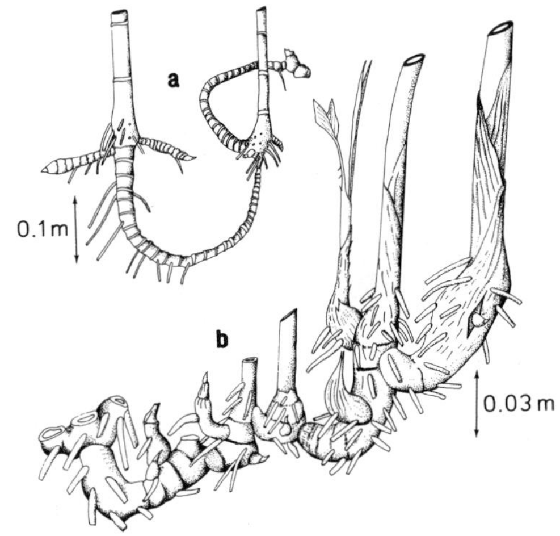

Fig. 22a,b. Rhizomes of small, multistemmed palms. **a** *Bactris elegans* (see Fig. 37). **b** *B. mitis*

multistemmed (Tomlinson's model), or to a true single-stemmed species (Corner's model). The palm is multistemmed when the individuals produce several stems during their lifetime and single-stemmed when they produce only one axis. Multistemmed palms are cespitose when they produced several to many contemporaneous stems, all issued from the base of the mother axis; they are rhizomatous when the new axes firstly develop a geotropic or horizontal phase of growth with scale leaves (Fig. 22) and then elaborate aerial stems with functional leaves; they are stoloniferous when the axes run along the ground and produce functional leaves during the horizontal growth phase, then the apex roots and develops an aerial stem usually larger in diameter than the stolon (Fig. 23).

Basal branching appears to be a specific character. In Arecoideae, all Amazonian species of *Bactris*, *Desmoncus*, *Prestoea*, and *Hyospathe elegans* are multistemmed palms. *Aiphanes*, *Astrocaryum*, *Euterpe*, *Geonoma*, *Leopoldinia*, *Manicaria*, *Oenocarpus*, *Syagrus* have multistemmed representatives. In Iriarteeae, *Iriartella setigera*, *I. stenocarpa* (Kahn 1983; Henderson 1990), *Catoblastus drudei*, *Socratea salazarii*, and the tall *Dictyocaryum ptariense* can produce stolons from the stem, usually from between the stilt roots. The stolons are abortive in the two latter species. This is not the case with *Iriartella* spp. and *C. drudei*, which form groups of adult axes; these are most often separated (Fig. 24) as a result of the degenerescence of the stolons after apex rooting and development into a stem. *Wettinia augusta* is also a multistemmed palm. *Lepidocaryum*, *Mauritiella*, and *Raphia* in Calamoideae, *Wendlandiella* and some species of *Chamaedorea* in Ceroxyloideae, *Ammandra* and *Phytelephas* in Phytelephantoideae, one species of *Chelyocarpus* in Coryphoideae, all

Fig. 23. Stolons of *Iriartella setigera*

these have the potential to branch basally. Of the 39 Amazonian palm genera, 23 (59.0%) include multistemmed species.

In large palms, the number of axes per clump is usually high, up to 25 as reported by Cavalcante (1974) for *Euterpe oleracea*. *Mauritiella aculeata* produces up to 15 axes per clump, several of which are mature, as observed throughout the basin. *Astrocaryum jauari* can also produce clonal patches of many axes from only one seed. Only the seedlings and successive acaulescent juveniles are solitary; De Steven (1986) studied in detail the demography of *O. mapora* in a Panamean forest and noted (p. 101): "An individual palm begins producing ramets in the late juvenile phase; over a 3-year period, less than 1% of juveniles <0.5 m tall produced ramets while 18% of juveniles >0.5 m produced new ramets. Only about 12% of small (1–3 m) immature palms lack one or more ramets and all large immature individuals have them."

More than half of the populations of most medium-sized, multistemmed species are solitary axes. For instance, the percentage represented by clustered individuals in *Oenocarpus minor* and *Wettinia augusta* populations is 21.2 and 25.6%, respectively; clumps are generally composed of one adult axis and one or several juveniles. In some species, clumps of many axes are rather frequent. For instance, *Syagrus stratincola* grows in dense clumps of 5 to 20 stems and produces many suckers. Clumps of *Astrocaryum carnosum* and *Oenocarpus mapora* are often composed of several adult and several juvenile axes. A higher percentage of axes in these populations are clumped,

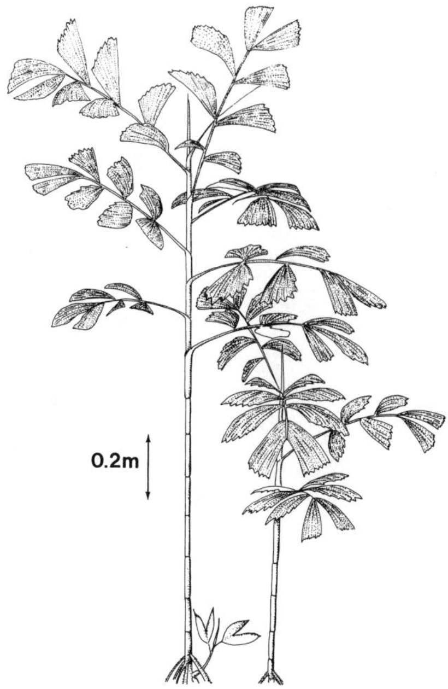

Fig. 24. *Iriartella stenocarpa*

43.0 and 49.2%, respectively (Table 10A,B). As with arborescent, multi-stemmed palms, the solitary axes are essentially seedlings and juveniles.

Most small species are multistemmed. In the populations of *Bactris elegans*, *B. sphaerocarpa* Trail, *Geonoma deversa*, *G. maxima*, the majority of axes form clumps; in the populations of *Astrocaryum gynacanthum*, *B. humilis*, *B. monticola*, *Geonoma piscicauda*, the number of solitary axes is higher (Table 11A). Clumps with two to four axes are the most frequent, representing 86.7% of all the clumps of small palms counted (Table 11B). The smallest species of *Bactris*, such as *B. simplicifrons*, produce only one adult axis with one juvenile at the base, or two contemporaneous adults, rarely more.

Only palms which develop Tomlinson's model are able to multiply by suckering, rooting, and fragmentation. This frequently occurs in rhizomatous or stoloniferous palms. *Bactris major* Jacquin, *B. campestris* Poeppig ex Martius, and *B. maraja* (Fig. 25) form clumps of many axes, the first along estuaries and beaches of the Atlantic coast, the second in waterlogged, white-sandy soils, the third in the sediments newly deposited by the rivers; the many erect stems are connected by a dense network of rhizomes creeping in the mud. The populations are generally composed of isolated clonal clumps from fragmentation of a single parent. Fragmentation in stoloniferous species is due to degenerescence of stolons, making solitary axes as a result. That has been observed in *Lepidocaryum tessmannii*, which is most often encountered as one axis, but a remnant part of a stolon is

Table 10. Basal branching in medium-sized palms

A) Population structure	No. of axes	
	Solitary	Clustered
Astrocaryum carnosum[a]	332 (57.0%)	250 (43.0%)
Oenocarpus mapora[b]	347 (50.8%)	336 (49.2%)
Oenocarpus minor[c]	63 (78.8%)	17 (21.2%)
Wettinia augusta[d]	227 (74.4%)	78 (25.6%)

B) Clump frequency in relation to number of axes per clump

	No. of axes per clump													
	2	3	4	5	6	7	8	9	10	11	12	13	17	18
Astrocaryum carnosum[a]	20	18	14	6	4	4	1	0	1	0	0	0	0	0
Oenocarpus mapora[b]	12	13	8	6	5	5	1	4	2	2	1	1	1	1
Oenocarpus minor[c]	5	1	1	0	0	0	0	0	0	0	0	0	0	0
Wettinia augusta[d]	23	6	2	0	1	0	0	0	0	0	0	0	0	0

[a] 0.40 ha.
[b] 3.2 ha.
[c] 0.96 ha.
[d] 2.8 ha surveyed.

Table 11. Basal branching in small palms

A) Population structure	No. of axes	
	Solitary	Clustered
Astrocaryum gynacanthum[a]	1867 (61.9%)	1151 (38.1%)
Bactris elegans[b]	113 (21.5%)	412 (78.5%)
Bactris humilis[b]	225 (80.9%)	53 (19.1%)
Bactris monticola[b]	102 (60.4%)	67 (39.6%)
Bactris sphaerocarpa[c]	364 (29.4%)	873 (70.6%)
Geonoma deversa[b]	87 (19.0%)	372 (81.0%)
Geonoma maxima[d]	17 (18.9%)	73 (81.1%)
Geonoma piscicauda[c]	126 (70.8%)	52 (29.2%)

B) Clump frequency in relation to number of axes per clump

	No. of axes per clump													
	2	3	4	5	6	7	8	9	10	11	12	13	17	18
Astrocaryum gynacanthum[a]	162	107	61	27	11	6	1	0	0	1	0	0	0	0
Bactris elegans[b]	44	34	25	8	6	4	1	0	1	0	0	0	0	0

[a] 10.56 ha.
[b] 3.84 ha.
[c] 0.71 ha.
[d] 6.0 ha surveyed.

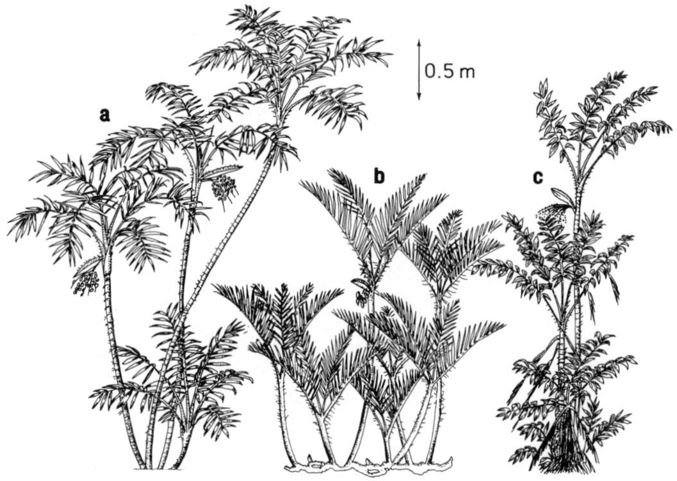

Fig. 25a–c. Small, multistemmed palms forming clumps of many axes. **a** *Bactris maraja*. **b** *B. major*. **c** *Bactris campestris*

always present at its base (Kahn and Mejia 1987), as well as in *Iriartella* spp. as described above. Rooting and fragmentation of stems within a clump has been observed in *Geonoma baculifera* (de Granville 1977), *G. deversa*, and *Hyospathe elegans*. The longest canes bend down to the ground; suckers with bundles of roots arise from axillary buds at the upper face of the horizontal canes and produce new individuals after rooting and fragmentation (Fig. 26).

2.2.3 Leaves

The leaves of Amazonian palms are palmate, costapalmate, or pinnate; these forms and those of the segments and pinnae are described in the supplement (Figs. 63 and 64). The blade is often entire, bifid at the tip, to irregularly dissected in a few broad segments in the smallest species (*Asterogyne*, *Bactris*, *Chamaedorea*, *Geonoma*, *Hyospathe*, *Wendlandiella*) and in the medium-sized *Manicaria*. Such an entire blade is considered as a juvenile character (Tomlinson 1979, 1990), and the species which develop it are interpreted as neotenic plants, i.e., sexually functioning with juvenile characters (Guillaumet 1973; de Granville 1977; see Sect. 5.2). The smallest leaves (20-cm-long, entire, bifid blade) are found in the dwarf, multistemmed palms (*Bactris aubletiana*, *B. simplicifrons*, *Geonoma stricta*, *G. leptospadix*, *G. pycnostachys*). The leaves of the subacaulescent species are entire, bifid, or pinnate, and usually larger, 0.7 m or more in length.

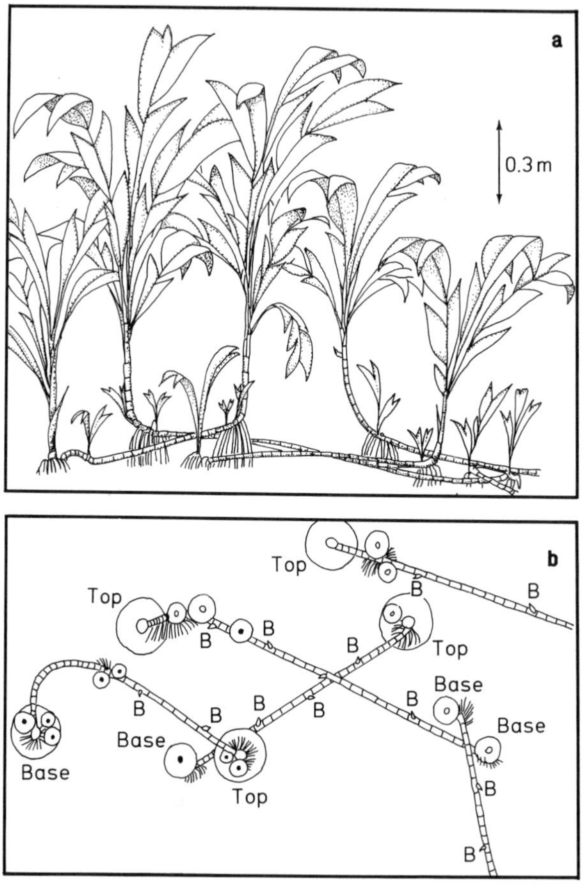

Fig. 26a,b. Rooting and fragmentation of the stem in *Geonoma baculifera*. **a** The longest canes bend down to the ground. **b** Suckers arise from axillary buds (*B*) and produce new individuals after rooting and fragmentation

Leaflets (or pinnae) of pinnate leaves are either regularly disposed in one plane along the rachis, or oriented in several directions and then, usually arranged in groups ("ragged" leaves). When both dispositions occur in a genus, the taller species generally present ragged leaves, while the pinnae of the medium-sized and small species are regularly disposed in one plane (e.g., *Astrocaryum chambira*, *A. jauari* vs. *A. macrocalyx* and *A. gynacanthum*; *Oenocarpus bacaba* vs. *O. minor*). This is also true with Amazonian Iriarteeae, the pinnae of which are usually entire in medium-sized (*Socratea salazarii*, *Wettinia augusta*, *W. maynensis*) and smaller species (*Catoblastus drudei*, *Iriartella setigera*, *I. stenocarpa*), and longitudinally divided in two to nine segments in larger species (*Dictyocaryum lamarckianum*, *D. ptariense*, *Iriartea deltoidea*, *Socratea exorrhiza*). These segments, which are oriented in several directions from the rachis, form a ragged leaf. The large leaves of most subterranean-stemmed palms have the pinnae regularly arranged in one plane. However, *Astrocaryum acaule* and *Scheelea insignis* develop very large leaves, the pinnae of which are disposed in groups of two to seven and oriented in several directions.

Leaf morphology usually varies during the ontogenesis. Leaves of seedlings of pinnate-leafed palms are entire, bifid or not (*Acrocomia*, *Astrocaryum*, *Attalea*, *Bactris*, *Catoblastus*, *Chamaedorea*, *Elaeis*, *Geonoma*, *Hyospathe*, *Iriartea*, *Maximiliana*, *Orbignya*, *Scheelea*, *Wendlandiella*, *Wettinia*), with only two pinnae (*Aiphanes*, *Desmoncus*, *Dictyocaryum*, *Euterpe oleracea*, *Geonoma*, *Jessenia*, *Oenocarpus*, *Socratea*), or three or more pinnae (*Bactris elegans*, *Euterpe precatoria*). Leaves of seedlings of palmate-leafed palms present two segments (*Chelyocarpus*, *Itaya*, *Lepidocaryum*, *Mauritiella*); those of costapalmate palms are palmate with three or more segments (*Mauritia*). In most species, pinnae of juveniles leaves are regularly disposed in one plane, while those of adult leaves are oriented in several directions. The transformation into a ragged leaf generally occurs when the palm is still acaulescent (*Astrocaryum aculeatum*, *A. chambira*, *Maximiliana maripa*, *Oenocarpus bacaba*, *O. distichus*, *Scheelea* spp.). The pinnae of *Iriartea deltoidea* start dividing longitudinally, and form the first ragged leaves while the palm has elaborated a several meter-high trunk.

A few Amazonian species are known to have a distichous crown: *Oenocarpus distichus*, *O. discolor*, and *O. tarampabo* (Balick 1986; see Fig. 16 in Balslev and Moraes 1989). In these cases, the distichous crown is a specific character and all palms have the leaves arranged in two ranks. Only one individual of *Mauritia flexuosa* with distichous crown was described near Iquitos (Kahn 1988a), and a second "abnormal" palm was found by K. Mejia (pers. comm. 1990) in the same area. Oldeman cultivated several distichous individuals of lianescent *Desmoncus* sp. in French Guiana (pers. comm. 1991).

Armed leaves are common in Bactridinae. The edges of the blade always bear setae in *Acrocomia*, *Astrocaryum*, and *Bactris*, but not in *Aiphanes* and *Desmoncus*. In most species of these genera, sheath, petiole, and rachis are strongly armed. However, the smallest species of *Bactris*, such as *Bactris aubletiana* and *B. simplicifons*, are unarmed, except for the setae at leaf edge. *Acrocomia*, the tallest species of *Astrocaryum*, some *Aiphanes*, *Bactris*, and *Desmoncus* develop spines on the blade. The largest spines, up to 30 cm in length, are produced on the sheath of *Astrocaryum javarense*, *A. huicungo*, *A. macrocalyx*, *A. murumuru*, and *A. urostachys*. *Desmoncus* are characterized by their leaves with a rachis prolonged into a cirrus armed with several pairs of recurved hooks. The leaves of Lepidocaryeae are armed with short spines usually swollen at the base; they are regularly disposed on the ribs (*Mauritia*) and on the margins of the blade (*Lepidocaryum*, *Raphia*). The petiole of *Elaeis* is armed with hooks which are modified pinnae.

2.2.4 Stems

2.2.4.1 Spines on Stem

Acrocomia, *Aiphanes*, *Astrocaryum*, *Bactris* usually present large and dense spines, often distributed in rings at leaf scars. The trunk of *Mauritiella aculeata* is armed with conical, sharp, up to 7-cm-long spines. These are roots which are produced on the trunk from the base to the crownshaft. Such spine-roots are occasionally found on the trunk of *Mauritia flexuosa* (de Granville 1974).

2.2.4.2 Swollen Stem

A few Amazonian palms present a swollen stem. The most famous is *Iriartea deltoidea* in the lowlands of western Amazonia, while this characteristic is not so well marked and less frequent in the populations found in the forests of the eastern slopes of the Andes. A swollen trunk is also found in *Astrocaryum chambira* when growing in an open area.

2.2.4.3 Muff of Dead Leaf Sheaths on Stem

Several medium-sized and small palms have the stem always covered by a muff of the sheaths of the dead leaves. In some cases, such as *Astrocaryum murumuru*, *Leopoldinia* spp., the muff covers the whole stem. In other cases, the sheaths of dead leaves persist only in the higher part of the stem under the crown, such as in *Astrocaryum macrocalyx*, *A. sciophilum*, *Phytelephas macrocarpa*, *Aphandra natalia*, and the tall palms of the complex *Attalea-Maximiliana-Orbignya-Scheelea*. In *Acrocomia lasiospatha*, the whole dead leaves remain hanging under the crown.

With a muff of dead leaf sheaths around the stem, the palm becomes the habitat of many animals, invertebrates and vertebrates. This fauna is particularly abundant and diversified when the sheath is fibrous or spiny (Chap. 6).

2.2.5 Roots

2.2.5.1 Stilt Roots

Iriarteeae produce stilt roots from their stem. These roots are remarkably developed by *Socratea exorrhiza* and *Iriartea deltoidea* (Wallace 1853; Bouillenne 1924; Kahn 1977, 1986b; Henderson 1990; Tomlinson 1990). In

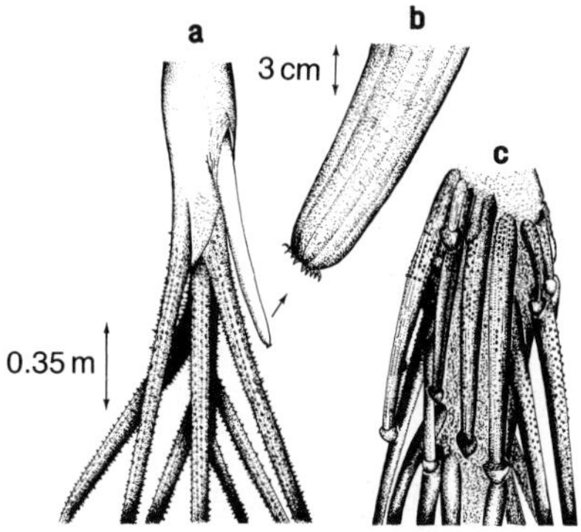

Fig. 27a–c. *Stilt roots:* **a,b** *Socratea exorrhiza.* **c** *Iriartea deltoidea*

the former species, the light-brown stilt roots form a cone up to 4 m in height, each root with a diameter of 4 to 7 cm, bearing white, horn-like spine-rootlets which appear after the apex of the stilt root enters the soil (Fig. 27a,b). Thus the stilt root is smooth during its growth phase, the apex bearing a small cap. The stilt roots of *Iriartea deltoidea* are black and form a rather dense closed cone, the white spine-rootlets are not so sharp, and a larger cap covers the apex during its growth from the stem to the ground (Fig. 27c). *Dictyocaryum lamarckianum* and *D. ptariense* also develop stilt roots. The cone of stilt roots is up to 50 cm high in medium-sized species, such as *Socratea salazarii* and *Wettinia* spp., and 10–20 cm in the smaller species, such as *Catoblastus drudei*, *Iriartella setigera*, and *I. stenocarpa*. Though their size is smaller, all the characteristics of Iriarteeae stilt roots can be recognized. Other palms present aerial roots at the base of their stem, such as *Asterogyne guianensis*, *Euterpe oleracea*, *Hyospathe elegans* (de Granville 1974; de Granville and Henderson 1988).

2.2.5.2 Pneumatophores

The root systems of plants which grow on waterlogged and flooded soils must work in anaerobic conditions. Data on the anatomy and physiology of such roots in palms have been given by Gatin (1907), Ernould (1921), and de Granville (1974), who studied the case of two Amazonian palms, *Mauritia flexuosa* (Fig. 28) and *Euterpe oleracea* (Fig. 29). Jost (1887) distinguished two kinds of aeriferous root structures in palms that he called "pneumathodes". The first is white, cone-shaped, and wart-like, and the other is whitish and floury, forming small plates covering the vertical root. De Granville (1974) called the former "pneumatorhizae" and the latter "pneumatozones". In Amazonian forests, pneumatorhizae are developed in *Astrocaryum*, *Attalea*, *Bactris*, *Catoblastus*, *Chamaedorea*, *Desmoncus*, *Dictyocaryum*, *Elaeis*, *Euterpe*, *Geonoma*, *Iriartea*, *Iriartella*, *Maximiliana*,

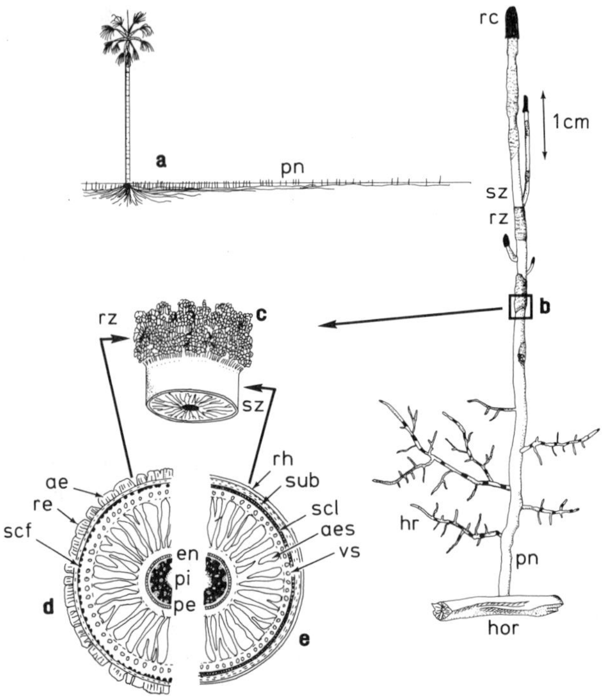

Fig. 28a–e. Pneumatophores of *Mauritia flexuosa*. **a** Palm and root system. **b** Pneumatophore. **c** Detail of a pneumatozone (rough zone). **d**,**e** Cross-sections in rough zone (**d**) and smooth zone (**e**). *ae* aerenchyma; *aes* aeriferous space; *en* endodermis; *hr* hair roots; *hor* horizontal root; *pe* pericycle; *pi* pith; *pn* pneumatophores; *rc* root cap; *re* remnant of rhizodermis; *rh* rhizodermis; *rz* rough zone; *scf* sclerified fibers; *scl* sclerenchyma; *sub* suberous sclerified fibers; *sz* smooth zone; *vs* vessels

Oenocarpus, *Orbignya*, *Scheelea*, *Socratea*, and *Wettinia*; pneumatozones are found in Lepidocaryeae, *Mauritia flexuosa* and *Raphia taedigera*. Oldeman (1971) explained these pneumatophores as a physiological complex of root axes. A root which grows vertically from the soil to the air produces on its aerial part many aeriferous structures, while its subterranean part branches into small, thin, absorbing roots. These latter are able to function with the oxygen obtained through the aeriferous structure.

2.2.5.3 Muff of Roots on Stem

Persistent bundles of roots located at variable heights around the stem are sometimes observed. These roots, which rarely reach the ground, are more frequently found in large, single stemmed palms, such as *Oenocarpus bacaba* and *Socratea exorrhiza*. An individual of this latter species had developed a second cone of stilt roots at several meters above the basal

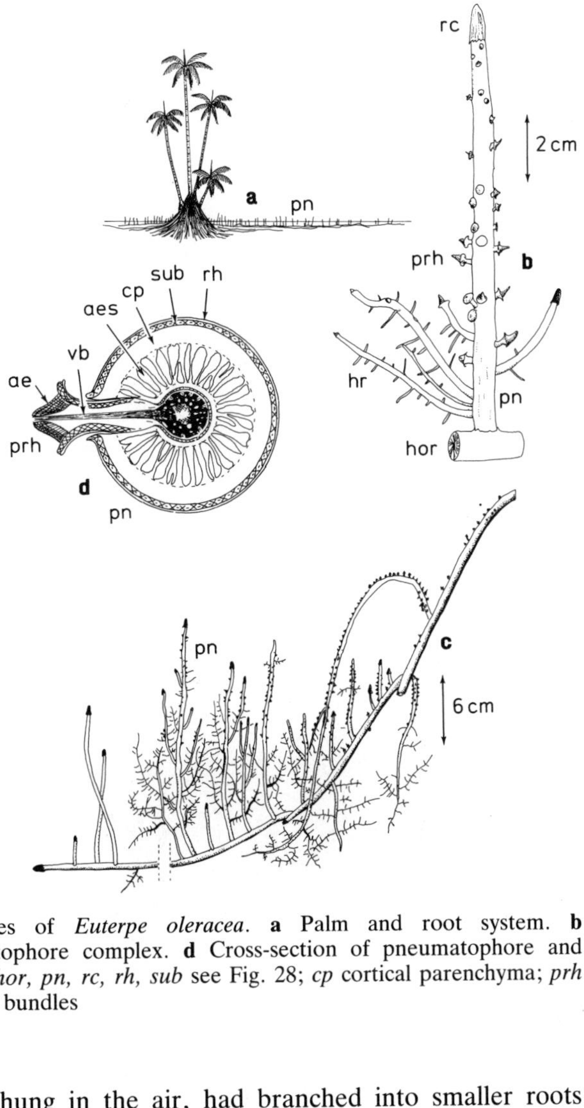

Fig. 29a–d. Pneumatophores of *Euterpe oleracea*. **a** Palm and root system. **b** Pneumatophore. **c** Pneumatophore complex. **d** Cross-section of pneumatophore and pneumatorhiza. *ae, aes, hr, hor, pn, rc, rh, sub* see Fig. 28; *cp* cortical parenchyma; *prh* pneumatorhizae; *vb* vascular bundles

cone; the roots, which hung in the air, had branched into smaller roots after a 2-m-length growth, as they commonly do when they enter the ground (Kahn 1983). The abnormal production of these stilt roots and their branching could not be explained by high flooding because the palm was located on a unflooded hill top. Another example of aerial muff of roots is provided by *Mauritia flexuosa* which can produce many short (2–10 cm long), erect rootlets on its stem.

3 Palm Communities in the Forest Ecosystems of Amazonia

3.1 Introduction

The perception that the jungle totally covers the Amazon basin is accurate. This does not mean, however, that the forest is uniform. In fact, there are several forest types, each with clearly distinct floristic composition and architecture (Pires 1978; Guillaumet 1987). Forests on three main geomorphological units will be considered: unflooded uplands which include drylands and waterlogged soil areas; wetlands which are irregularly, periodically, or permanently flooded; and montane forests at medium and high elevation located on eastern Andean slopes, up to 2500 m, and on a few reliefs scattered at the periphery of the Amazon basin, in Guyana and southern Venezuela (complex of old sandstone table mountains, up to 3000 m in altitude, commonly known as "Guayana highland" or "tepuis" in Venezuela). The palm communities in these ecosystems will be characterized by their diversity of species and life forms, the density and vertical distribution of each species. An overview of Amazonian ecosystems is incomplete if it fails to take into account deforested areas, which have been rapidly increasing for the last 10 years, savannahs, and inselbergs which emerge from the green ocean of forests. Such an attitude was assumed by Daly and Prance (1989, p. 406) in their presentation of Brazilian Amazon forests, "some non-forested areas are included because they are highly specialized formations which have unique floras, are of limited expanse, and are situated well within the limits of the classic hylaean forests".

3.2 Methods

For each inventory presented, all palms were counted in delimited areas. In the case of multistemmed species, each axis was counted as a separate plant.

All species have been collected; voucher numbers and herbaria are cited in the respective publications. Some species were not identified because of the lack of either fertile material or of comparable specimens in the herbaria. Distinguishing palm species in the field was possible after assembling a collection from seedling to adult of each species involved and working with palms in these forests for years.

3.3 Unflooded Forests

The Amazonian concept of "terra firme" (Brazil), "tierra firme" (Bolivia, Ecuador, Venezuela), "terre ferme" (French Guiana), "droogland" (Surinam), or "altura" (Peru) includes all lands which are never flooded by rivers or small streams. Various ecosystems stand on these lands: terra firme forests on usually well-drained, or sometimes poorly drained soils, shrubby vegetation on dry white sand, palm-dominant forests on white sand water-logged after rainfall.

3.3.1 Terra Firme Forests

"Terra firme forest", sensu Prance (1978), covers plateaus and valley slopes. Soils are most often clayey and well drained; they correspond to acrisol and luvisol in the FAO-UNESCO (1971) classification system. This forest type, dominated by large trees, is characterized by having the best-developed vertical architecture. Medium-sized trees and shrubs of different adult heights account for the noticeable complexity of this ecosystem.

Terra firme forests are also characterized by their very high diversity in tree species. In central Amazonia near Manaus, Prance et al. (1976) counted 179 tree species (DBH > 15 cm) on 1 ha, belonging to 43 families, ten of them with more than ten species each. In eastern Amazonia, Black et al. (1950), Pires et al. (1953), and Pires (1978) found 87 species (DBH > 10 cm) on 1 ha, 179 species (DBH > 10 cm) on 3.5 ha, and 224 species (DBH > 30 cm) on 5 ha surveyed, respectively, and Campbell et al. (1986) listed 118 to 162 species (DBH > 10 cm) in three 1-ha samples along the Rio Xingú. In French Guiana, Sabatier and Prevost (1989) inventoried 130 and 175 tree species (DBH > 10 cm) on 0.5 and 1 ha surveyed respectively at Piste de Saint Elie, and Mori and Boom (1987) reported 293 species for a sample of 800 trees (DBH > 10 cm) along a 4-km-long track near Saül. In western Amazonia, Gentry (1986a) collected up to 249 species on 0.1 ha near Iquitos, and Balslev et al. (1987) estimated the species richness in Ecuadorean Amazonia to be up to 228 species/ha. In the southern region of the basin, in Bolivia, Boom (1986) inventoried 94 tree species (DBH > 10 cm) on 1 ha surveyed. Gentry (1982) gathered together most inventory carried out in tropical rain forests in neotropics; all contribute to point out the very high diversity of these forests.

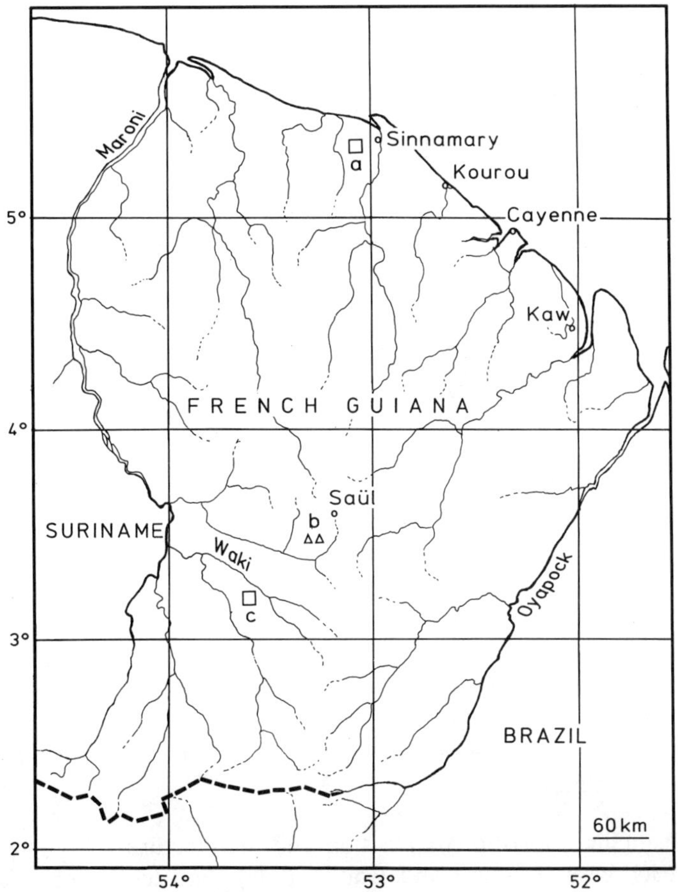

Fig. 30. French Guiana. *a* Piste de Saint Elie; *b* Galbao mountains; *c* lower Waki River valley

3.3.1.1 Palms in the Terra Firme Forests in Eastern Amazonia

French Guiana. The coastal forest surveyed by Sist (1989a) extends from km 13 to km 16 of the road called "Piste de Saint Elie" (Fig. 30a), South to Sinnamari village (5°00′N; 53°00′W). The climate is humid tropical. Average annual rainfall is 3100 mm with a peak during April and June. The rainy season, from December to July, is interrupted in March by a short dry season ("the small March summer"). The dry season extends from August to October. Relative humidity in the understory ranges from 89 to 96% between 16.00 h and 11.00 h and drops to 66% around 14.00 h. Two soil classes are recognized in the uplands according to water dynamics (Boulet et al. 1979): the soils with a free vertical drainage which are never waterlogged; and those with a bad vertical drainage which are waterlogged after

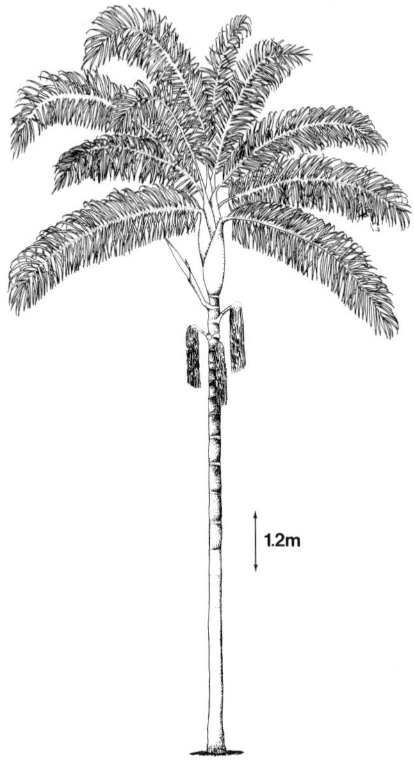

Fig. 31. *Oenocarpus bacaba*

rainfalls. Floristic composition and architecture is significantly different in each case (Lescure and Boulet 1985). Fourteen species and eight genera were found on 0.5 ha on well-drained soils (Table 12A). Four species, *Astrocaryum sciophilum*, *Bactris aubletiana*, *Jessenia bataua* subsp. *oligocarpa*, and *Oenocarpus bacaba* (Fig. 31) represent 87.6% of the community. Density is 143 palms/0.1 ha. Thirteen species and seven genera were found on a 0.25-ha plot on soils with bad vertical drainage (Table 12B). Three species, *Astrocaryum sciophilum*, *Jessenia bataua* subsp. *oligocarpa*, and *Oenocarpus bacaba* represent 80.8% of the community. Density is 173 palms/0.1 ha. In both forests, *Euterpe oleracea* is only present as seedlings.

Palms have been inventoried along a transect in the terra firme forests of the Galbao mountains, the summits of which culminate at 660 and 720 m above sea level (de Granville 1989). They are located in the central part of French Guiana (3°35′N, 53°18′W), 10 km west of Saül (Fig. 30b), on a basic bedrock (gabbros). The climate is humid tropical with an average annual rainfalls. Floristic composition and architecture is significantly different in each case (Lescure and Boulet 1985). Fourteen species and eight genera were found on 0.5 ha on well-drained soils (Table 12A). Four

Table 12. Species richness, density, and vertical distribution of palms in a coastal terra firme forest at Piste de Saint Elie, French Guiana (Sist 1989a)

	Seedlings	Juveniles and adults	Total
A) In 0.5 ha on well-drained soils in French Guiana			
Astrocaryum paramaca	1	3	4
Astrocaryum sciophilum	144	101	245
Bactris acanthocarpoides	11	10	21
Bactris aubletiana	104	51	155
Bactris constanciae	2	3	5
Bactris gastoniana	3	9	12
Bactris oligocarpa	3	9	12
Bactris rhaphidacantha	3	3	6
Euterpe oleracea	6	0	6
Jessenia bataua[a]	64	70	134
Maximiliana maripa	1	1	2
Oenocarpus bacaba	86	8	94
Scheelea sp.	6	13	19
Socratea exorrhiza	2	0	2
Total	436	281	717
B) In 0.25 ha on poorly drained soils			
Astrocaryum paramaca	0	6	6
Astrocaryum sciophilum	86	56	142
Bactris acanthocarpoides	0	2	2
Bactris aubletiana	32	9	41
Bactris gastoniana	1	1	2
Bactris oligocarpa	2	2	4
Bactris rhaphidacantha	3	3	6
Euterpe oleracea	5	0	5
Geonoma oldemanii	1	2	3
Geonoma maxima	0	1	1
Jessenia bataua[a]	39	57	96
Oenocarpus bacaba	108	3	111
Scheelea sp.	10	3	13
Total	287	145	432

[a] *Jessenia bataua* subsp. *oligocarpa*.

a canopy dominated by Sapotaceae, Leguminosae, and Lecythidaceae. Highest trees seldom exceed 40 m in height. Treefalls are more frequent on crests where trees are more exposed to wind. The survey area includes 63, 0.02-ha plots distributed along a 1.3-km strip from 320 to 660 m above sea level. Floristic composition varies along the transect. Tree diversity decreases with increasing elevation. Some species are found only in the lowest part of the transect, such as *Terminalia amazonia* (Combretaceae) and *Minquartia guianensis* (Olacaceae), while others occur only in the highest part, such as *Newtonia* sp. (Mimosaceae). On the summit, persistent mist promotes epiphytic life. Ten species and six genera of palms were

Table 13. Species richness, density, and vertical distribution of palms in inland terra firme forests, French Guiana

	h < 10 m	h > 10 m
A) Along a transect (1.26 ha) at Galbao mountains		
Astrocaryum paramaca	54	0
Bactris gastoniana	191	0
Bactris rhaphidacantha	39	0
Bactris simplicifrons	12	0
Geonoma euspatha	1	0
Geonoma maxima	10	0
Geonoma triglochin	146	0
Hyospathe elegans	196	0
Oenocarpus bacaba	204	1
Socratea exorrhiza	10	0
Total	863	1
B) In a 0.18-ha plot in the lower Waki River valley, French Guiana		
Astrocaryum paramaca	256	0
Bactris gastoniana	25	0
Bactris rhaphidacantha	54	0
Bactris simplicifrons	3	0
B. sp. aff. *B. acanthocarpoides*	3	0
Euterpe oleracea	3	0
Oenocarpus bacaba	27	2
Socratea exorrhiza	1	2
Syagrus inajai	1	0
Total	373	4

found on the slopes of the Galbao mountains on 1.26 ha surveyed. Mean density of palms was 68.6/0.1 ha; *Oenocarpus bacaba* mainly represented by seedlings and juveniles – only one palm with a height over 10 m was counted – *Hyospathe elegans*, *Bactris gastoniana*, and *Geonoma triglochin* are the dominant species (Table 13A). Two species, *Geonoma euspatha* and *G. triglochin*, were only found on the summit (Fig. 32); the density of the latter was particularly high with up to 64 individuals per 0.02-ha plots. Most species were present along the transect; some of them formed dense patches, such as *Hyospathe elegans* above 400 m, *Bactris gastoniana* and *Oenocarpus bacaba* in the lower half of the transect, and *Astrocaryum paramaca* at the lowest elevation. *Bactris rhaphidacantha*, *B. simplicifrons*, *Geonoma maxima*, and *Socratea exorrhiza* were scarcely found along the transect.

Palms have been also inventoried in a terra firme forest of the lower Waki River valley (3°19′N, 53°40′W) which is located at 50 km west of Saül (Fig. 30c) at low elevation, 130–150 m. *Dicorynia guianensis* (Papilionaceae) is the dominant species, with two individuals reaching 35 and 40 m in height

Fig. 32. *Geonoma triglochin* at the summit of Galbao mountains (660 m in elevation)

respectively, and with 11 trees of the 46 found over a height of 10 m. The other trees include two other species of Papilionaceae, and the following families, each represented by one or two species: Annonaceae, Apocynaceae, Burseraceae, Cecropiaceae, Chrysobalanaceae, Elaeocarpaceae, Lauraceae, Lecythidaceae, Mimosaceae, Moraceae, Myristicaceae, Myrtaceae, Sapotaceae, and Sterculiaceae. The density of treelets is low in the understory, which is dominated by palms. Nine species in six genera were found in 0.18 ha surveyed (180 × 10 m). Mean density of palms is 209/0.1 ha (Table 13 B). *Astrocaryum paramaca*, a subterranean-stemmed palm, is particularly abundant, and represents 67.9% of the community (Fig. 33). Two individuals of *Socratea exorrhiza* and two of *Oenocarpus bacaba* were counted above 10 m in height. *Euterpe oleracea* in only represented by two seedlings and one juvenile. This species is particularly abundant in a neighboring seasonal swamp forest.

Surinam. All palms with a height over 1.5 m were inventoried on six 0.25-ha plots in terra firme forests (Van der Steege 1983). Nine species and seven genera were found on the total 1.5 ha surveyed. These are *Astrocaryum paramaca*, the dominant species, *A. sciophilum*, *Bactris* (two species), *Maximiliana maripa*, *Oenocarpus bacaba*, *Socratea exorrhiza*, *Euterpe oleracea*, and *Geonoma baculifera*. The presence of the two latter together in two plots indicates that these were likely to include some areas on waterlogged soils.

Fig. 33. Dense population of *Astrocaryum paramaca* in a terra firme forest in the lower Waki River valley

Pará, Brazil. The climate of the lower Tocantins River valley is of the humid tropical type, with 2500 mm average annual rainfall and 26 °C average annual temperature. The rainy season extends from November to June; the dryer period is in September. The soils are classified as "yellow red latosols" with low fertility and a low level of saturation of bases (Ranzani and Podesta 1982). The forests between Tucuruí and Itupiranga are dominated by the Brazil nut tree, *Bertholletia excelsa* (Lecythidaceae). This species, which is locally called "castanha do Pará", gives its name to the forests it dominates ("castanhais"). It is a very large tree, up to 50 m in height and 2.6 m in diameter; 24 individuals with DBH > 1.4 m were found on 9.6 ha surveyed (Kahn 1983).

In Tucuruí forest, the survey area (Fig. 34a) includes plateau, crest, slope and a depression with a central gully, 2–3 m wide, with a few isolated spots of poorly drained soils. The average basal area for trees with DBH > 1.5 cm was 35.1 m^2. A total of 214 trees (DBH > 40 cm) were counted representing a basal area of 19.7 m^2/ha for the 3.84 ha studied (Table 14). Twelve species and eight genera of palms were found with a total of 3446 individuals (axes

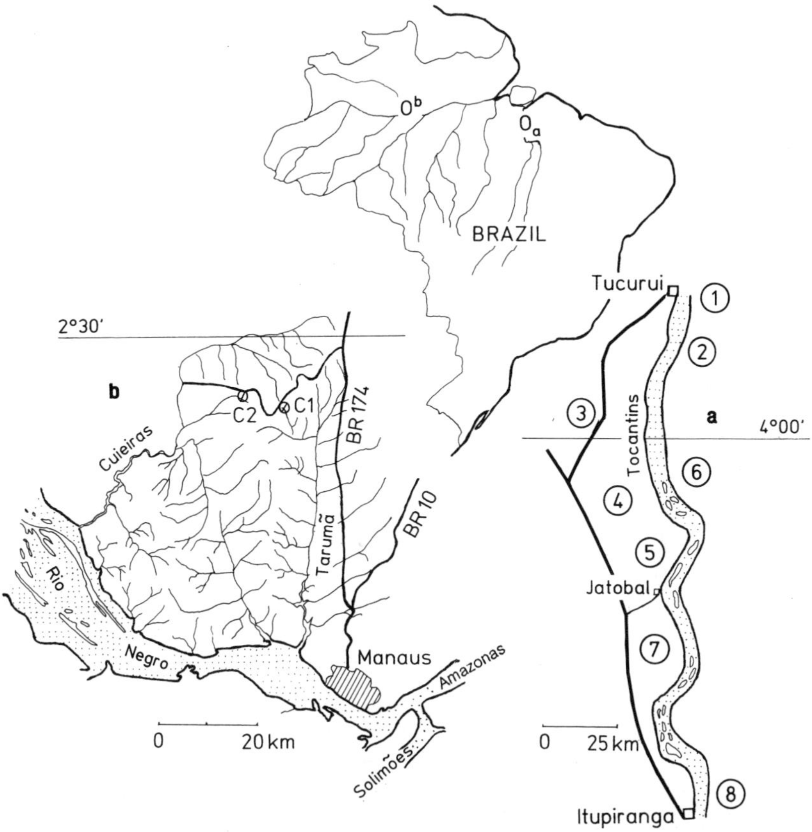

Fig. 34a,b. Map of Brazil with enlargements of the two areas studied in Amazonia. **a** Lower Tocantins River valley (survey areas are noted *1–8*, see Table 17). **b** Lower Rio Negro valley (survey areas are noted *c1–c2*)

for multistemmed species) on this area (Table 15). The mean density is 90/0.1 ha. The dominant life form in the understory is the multistemmed palm (72.3% of the palm community) mainly developed by *Astrocaryum gynacanthum* (34.3%), *Bactris elegans* (15.2%), and *Geonoma deversa* (13.3%). Adult arborescent palms (greater than 10 m in height) were poorly represented; only three individuals of *Oenocarpus bacaba* and one of *Socratea exorrhiza* were found. Adult plants of *Maximiliana maripa* were not encountered. Seedlings and juveniles of these three arborescent species represent, however, 25.2% of the community. *Euterpe oleracea*, with only four juveniles, was limited to the infrequent spots of poorly drained soils in the gullies in depressions.

Frequency and density of the former palms species (given on 3.84 ha) are now compared with data from seven other areas located along the lower Tocantins River valley between Tucuruí and Itupiranga. In each area,

Table 14. Family composition and relative importance of basal area (for DBH > 0.4 m) in 3.84 ha of a terra firme forest in the lower Tocantins River valley, near Tucuruí, Brazil. (Kahn 1987)

	No. of trees (DBH > 40 cm)	Basal area (%)
Leguminosae	76	35.3
Lecythidaceae	34	30.1
Chrysobalanaceae	49	12.8
Burseraceae	16	5.7
Anacardiaceae	6	4.8
Lauraceae	6	2.0
Sapotaceae	8	2.0
Meliaceae	4	1.3
Bignoniaceae	1	1.3
Myristicaceae	2	1.1
Tiliaceae	3	<1
Flacourtiaceae	3	<1
Melastomataceae	1	<1
Nyctaginaceae	2	<1
Bombacaceae	1	<1
Humiriaceae	1	<1
Sterculiaceae	1	<1

Table 15. Species richness, density, and vertical distribution of palms in 3.84 ha of terra firme forest in the lower Tocantins valley, near Tucuruí, Brazil. (Kahn 1987)

	Height classes			
	0–1 m	1–10 m	>10 m	Total
Astrocaryum gynacanthum	322	860	0	1182
Bactris elegans	150	375	0	525
Bactris humilis	56	222	0	278
Bactris simplicifrons	0	5	0	5
Bactris sp.	12	20	0	32
Euterpe oleracea	0	4	0	4
Geonoma deversa	155	304	0	459
Geonoma maxima	3	6	0	9
Maximiliana maripa	58	123	0	181
Oenocarpus bacaba	357	170	3	530
Socratea exorrhiza	53	107	1	161
Syagrus inajai	17	63	0	80
Total	1183	2259	4	3446

data are given for 0.72 ha. *Maximiliana maripa* (Fig. 35), *Astrocaryum gynacanthum* (Fig. 36), and *Bactris humilis* were found in all areas surveyed, and *Syagrus inajai* was absent in only one area. *Oenocarpus bacaba* and *O. distichus* were frequent in the northern and southern part of the valley, respectively. *Bactris* sp., *Geonoma maxima*, *Socratea exorrhiza* were found

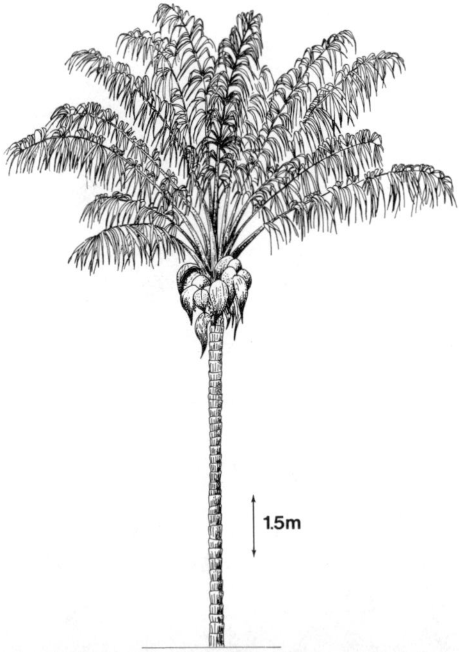

Fig. 35. *Maximiliana maripa*

Fig. 36. *Astrocaryum gynacanthum*, 5 m in height

Fig. 37. *Bactris elegans*, 3 m in height

in four of the eight areas surveyed. *Bactris elegans* (Fig. 37), which is very abundant in the former inventory, *Bactris simplicifrons* and *Geonoma deversa*, which are found throughout the Amazon basin, present a low frequency (Table 16). *Astrocaryum gynacanthum*, *Maximiliana maripa*, *Bactris humilis*, and *Syagrus inajai*, the most widely distributed in the lower Tocantins River valley, also form dense populations in these forests. *Astrocaryum gynacanthum*, in particular, is the most abundant palm in the understory. Seedlings and juveniles of *Maximiliana maripa* are found in high density, while only two individuals greater than 10 m in height were observed on the total 8.88 ha surveyed (Table 17).

3.3.1.2 Palms in a Terra Firme Forest in Central Amazonia

The survey area was located in the Tropical Silviculture Experimental Station of INPA, 45 km from Manaus (Fig. 34b), between the BR-174 (Manaus-Boa Vista Highway) and the Rio Cuieiras (2°35′–2°40′S; 60°00′–60°20′W). Rainfall is about 2500 mm per year. The rainy season extends from December to May, the dry season from June to November. The water deficit is low or nil and the potential evapotranspiration is regularly dis-

Table 16. Frequency of palms in the terra firme forests along the lower Tocantins valley, Brazil (from Tucuruí to Itupiranga, see Fig. 34a)

	1[a]	2[b]	3[b]	4[b]	5[b]	6[b]	7[b]	8[b]
Astrocaryum gynacanthum	+	+	+	+	+	+	+	+
Bactris elegans	+	–	–	–	+	+	–	–
Bactris humilis	+	+	+	+	+	+	+	+
Bactris simplicifrons	+	–	–	–	–	+	+	–
Bactris sp.	+	+	+	+	–	–	–	–
Geonoma deversa	+	–	–	–	–	–	–	–
Geonoma maxima	+	+	–	–	–	+	–	+
Maximiliana maripa	+	+	+	+	+	+	+	+
Oenocarpus bacaba	+	+	+	+	+	–	–	–
Oenocarpus distichus	–	–	–	+	–	+	+	+
Socratea exorrhiza	+	+	–	–	+	+	–	–
Syagrus inajai	+	+	+	+	+	+	–	+

[a] 3.84 ha.
[b] 0.72 ha surveyed.

Table 17. Density of the four most frequent palm species in the forests of the lower Tocantins River valley, Brazil. (After Kahn 1987)

	West margin Height classes					East margin Height classes			
	−1 m	1–10 m	>10 m	Total		−1 m	1–10 m	>10 m	Total
Astrocaryum gynacanthum									
(3)	26	143	0	169	(1)[a]	60	161	0	221
(4)	155	71	0	226	(2)	48	248	0	296
(5)	99	72	0	171	(6)	202	124	0	326
(7)	39	161	0	200	(8)	50	159	0	209
Bactris humilis									
(3)	5	27	0	32	(1)[a]	10	42	0	52
(4)	3	16	0	21	(2)	3	63	0	66
(5)	5	13	0	18	(6)	0	9	0	9
(7)	1	25	0	26	(8)	3	20	0	23
Syagrus inajai									
(3)	91	114	0	205	(1)[a]	3	12	0	15
(4)	1	19	0	20	(2)	1	31	0	32
(5)	2	3	0	5	(6)	2	10	0	12
(7)	0	0	0	0	(8)	0	12	0	12
Maximiliana maripa									
(3)	20	177	1	198	(1)[a]	11	23	0	34
(4)	34	77	0	111	(2)	20	178	1	199
(5)	0	6	0	6	(6)	2	17	0	19
(7)	0	10	0	10	(8)	0	3	0	3

[a] Mean for 0.72 ha calculated from 3.84 ha surveyed; the other data from 0.72 ha surveyed. The number of the survey area is given in parentheses (see Fig. 34a).

Table 18. Species richness, density, and vertical distribution of palms in a terra firme forest (0.72 ha) on the lower Rio Negro valley, near Manaus, Brazil. (Kahn and Castro 1985)

	Height classes			
	0–1 m	1–10 m	>10 m	Total
Astrocaryum gynacanthum	38	108	0	146
Astrocaryum sociale	122	444	0	566
Bactris (11 species)	84	135	0	198
Euterpe precatoria	49	11	0	60
Geonoma (6 species)	127	130	0	257
Iriartella setigera	25	7	0	32
Oenocarpus bacaba	476	167	4	647
Oenocarpus minor	4	56	0	60
Scheelea sp.	20	232	0	252
Socratea exorrhiza	13	5	0	18
Syagrus inajai	18	72	0	90
Total	976	1346	4	2326

tributed throughout the year (Ribeiro 1976). The soils are heavy-clayey on the plateau, sandy-clayey on crest and slope, and well drained in both cases (Ranzani 1980; Chauvel 1982).

According to Prance et al. (1976)'s data from a forest located at 30 km North of Manaus, the most diversified tree families on 1 ha are: Lecythidaceae (18 species-4 genera), Moraceae (15 spp.-10 gen.), Sapotaceae (14 spp.-9 gen.), Caesalpiniaceae (12 spp.-5 gen.), Burseraceae (12 spp.-2 gen.), Annonaceae (11 spp.-6 gen.), Chrysobalanaceae (11 spp.-4 gen.), Mimosaceae (9 spp.-4 gen.), Lauraceae (8 spp.-4 gen.), Vochysiaceae (8 spp.-3 gen.), Melastomataceae (8 spp.-2 gen.), Rubiaceae (7 spp.-7 gen.).

Palms were inventoried on 0.12-ha plots established on plateau, crest, and slope of two catenas, 6 km apart. The data are lumped because of a lack of distinctiveness between plots of each topographic site (Table 18). On 0.72 ha, 26 species and 9 genera were found. The genera *Bactris* and *Geonoma* are particularly diversified with 11 and 6 species, respectively. Palm density is high with 2326 palms (323/0.1 ha) counted in the area surveyed. The density of adult, arborescent palms is very low; only four individuals of *Oenocarpus bacaba* were encountered with a height greater than 10 m, though this species is the most represented with 27.8% of the community, all seedlings and juveniles. In the understory, the dominant life form is the subterranean-stemmed, large-leafed palms, represented by *Astrocaryum sociale* and *Scheelea* spec. nov. with 24.3 and 10.8% of the community, respectively.

In the same region, Klinge and Rodrigues (1971) counted 11 species and 196 individuals with a height over 1.5 m on 0.2 ha.

3.3.1.3 Palms in the Terra Firme Forests in Western Amazonia

In the Lower Ucayali River Basin, Peru. The two forests studied are located near Jenaro Herrera, 4°55′S; 73°35′–73°40′W (Fig. 38a). The physiography of this region was presented by López Parodi and Freitas (1990). The first site, called Ferrocaño, is about 6 km North of the field station of the Research Institute for Peruvian Amazonia (IIAP); the second, called Copal, is at km 15 of the only road. Both sites belong to the same geographic unit at an elevation of about 160 m above sea level. Climate is humid tropical, with an average annual rainfall of 2900 mm and average annual temperature of 26 °C. Soils of both sites are yellow and clayey in their upper part, brownish to dark brown at greater depth (acrisol); the upper 4 m are never waterlogged; the drainage is vertical. At the Copal site, the soil differs slightly, having a silty layer in the upper horizon at the lowest part of the slope, and a sandy-clayey texture at some places on the plateau at the margin of the area surveyed, corresponding to a zone of transition into an area of gleyic podzol which extends about 50 m beyond.

Data on floristic composition and architecture of the forest have been given by Marmillod (1982). Basal area is 26.9 m^2/ha (for DBH > 10 cm), and mainly due to the following families: Leguminosae (13.2%), Lecythidaceae (12.8%), Sapotaceae (11.3%), Myristicaceae (7.5%), Moraceae (6.8%), Lauraceae (5.5%), Chrysobalanaceae (4.9%), Vochysiaceae (4.8%), Palmae (3.1%), Burseraceae (2.6%), Humiriaceae (2.5%), Melastomataceae (2.1%), Apocynaceae (2.0%).

All palms were counted on contiguous 0.01-ha plots (20 × 5 m). Total areas surveyed at the Ferrocaño and Copal sites were 0.71 ha and 0.5 ha, respectively. Density values for each species are lumped for the whole area at each site (Kahn and Mejia 1991).

At the Ferrocaño site, 29 species in 16 genera were encountered in 0.71 ha (Table 19). The genus *Bactris* is the most diversified with ten species. Palm diversity varies from 7 to 14 species and five to nine genera per 0.01-ha plot. Palm density is very high (986/0.1 ha). The palm community is largely dominated by two small, multistemmed, understory species, *Lepidocaryum tessmannii* (Fig. 39) and *Bactris sphaerocarpa*, which represent, respectively, 54.3% and 17.7% of the community. Marmillod (1982) showed that the crowns of small understory palms in the 0.9–6.5 m strata represent 60 to 70% of the vegetal cover in these forests.

At the Copal site, 34 species in 21 genera were encountered in 0.5 ha surveyed (Table 20). *Geonoma* is the most diversified genus, with eight species. Palm diversity varies from 6 to 18 species and 5 to 13 genera per 0.01-ha plot. Of the 50 plots surveyed, 27 present 12 or more species, and 31 have 9 or more genera. Palm density is slightly lower than in the former site (768 palms per 0.1 ha). *Lepidocaryum tessmannii* represents only 7.4% of the community. *Astrocaryum* cf. *chonta* and *Phytelephas macrocarpa* were found only on the lowest part of the slope where the soil has a

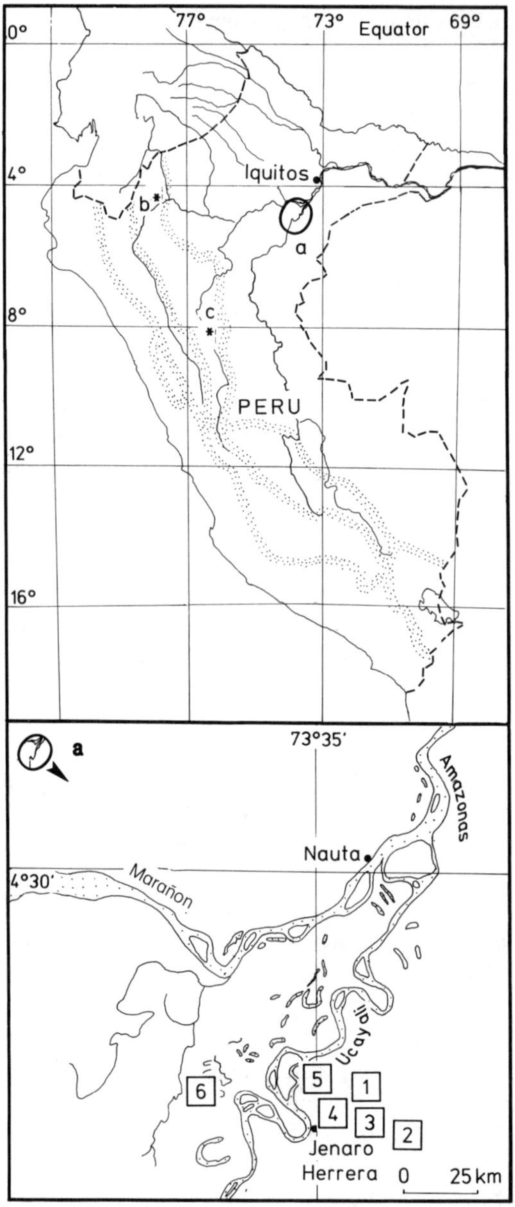

Fig. 38. Map of Peru with the three areas studied (*a*,*b*,*c*) in Amazonia. *a* lower Ucayali River valley, with enlargement (*bottom*) of Jenaro Herrera region (*1*,*2* terra firme forest at Ferrocaño and Copal site, respectively; *3* forest on white-sandy soils; *4* seasonal swamp forest; *5* forest on periodically flooded alluvial soils; *6 Mauritia flexuosa* swamp). *b* Santiago mouth; *c* upper Huallaga River valley

Table 19. Species richness, density, and vertical distribution of palms in 0.71 ha of terra firme forest at Ferrocaño site, near Jenaro Herrera, lower Ucayali River valley, Peru. (Kahn and Mejia 1991)

	Height classes			
	0–1 m	1–10 m	>10 m	Total
Astrocaryum chambira	0	1	0	1
Bactris acanthocarpoides	2	2	0	4
Bactris humilis	0	1	0	1
Bactris mitis	2	10	0	12
Bactris pectinata	3	11	0	14
Bactris piranga	0	1	0	1
Bactris simplicifrons	1	2	0	3
Bactris sphaerocarpa	339	898	0	1237
Bactris sp. aff. *B. mitis*	2	3	0	5
Bactris sp. 1	9	14	0	23
Bactris sp. 2	12	101	0	113
Chelyocarpus repens	18	188	0	206
Desmoncus cf. *leptospadix*	1	1	0	2
Euterpe precatoria	28	8	0	36
Geonoma acaulis	17	53	0	70
Geonoma leptospadix	15	22	0	37
Geonoma piscicauda	76	102	0	178
Geonoma poeppigiana	22	80	0	102
Geonoma spixiana	57	191	0	248
Hyospathe elegans	5	15	0	20
Iriartella stenocarpa	120	67	0	187
Jessenia bataua[a]	151	195	0	346
Lepidocaryum tessmannii	455	3346	0	3801
Maximiliana maripa	0	1	0	1
Oenocarpus balickii	6	3	4	13
Orbignya polysticha	8	182	0	190
Pholidostachys synanthera	27	117	0	144
Socratea exorrhiza	2	0	1	3
Wettinia augusta	1	5	0	6
Total	1379	5620	5	7004

[a] *Jessenia bataua* subsp. *bataua*.

distinctive silty layer in the upper horizon; both species are frequent on alluvial soils. The four adults of *Jessenia bataua* were located at the limit with a gleyic podzol, on which this species forms a dense population beyond the area surveyed.

At both sites, most of the species are small or medium-sized, understory palms, and most of the individuals of the palm community are less than 10 m in height. The density of large palms (greater than 10 m in height) is higher at Copal, with 25 individuals in 0.5 ha, than at Ferrocaño with only five in 0.71 ha. In this latter site, however, Marmillod (1982) found that palms represented 3.1% of the basal area. This appearent contradiction is due to

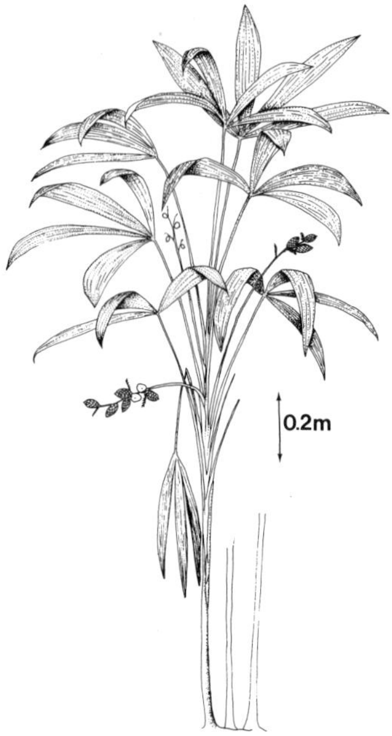

Fig. 39. *Lepidocaryum tessmannii*

the sample method. The 0.25-ha plots used by this author did not permit him to take into account the slight and short soil variations between acrisol and podzol. Kahn and Mejia (1991) disposed the 0.01-ha plots in order to avoid the areas of gleyic podzol; they were guided by a soil specialist (Luc Veillon, ORSTOM) using his auger. Thus, it was verified that large palms are in very low density on acrisol while they form dense population on gleyic podzol, as will be seen below. The percentage of palms under 1 m in height is low at Ferrocaño (19.7%) and high at Copal (62.4%). For the first site, the low value is the result of the very high density reached by *Lepidocaryum tessmannii* and *Bactris sphaerocarpa* between 1 and 10 m. For the second site, the high value is explained by the high density of seedlings of two arborescent species, *Iriartea deltoidea* and *Jessenia bataua*, and of juveniles of three small, understory species, *Hyospathe elegans*, *Chamaedorea pauciflora*, and *Geonoma piscicauda*, which are particularly abundant in the clearings.

In the Upper Marañon River Basin, Peru. Palms in the upland forest located at the mouth of Rio Santiago (Fig. 38b) are also very diversified. The following list of 23 species in 14 genera was obtained after a 4-h visit (F. Kahn and F. Borchsenius, pers. observ. May 1990): *Aiphanes deltoidea*, *A. tessmannii*, *A.* spec. nov., *Astrocaryum chambira*, *Bactris simplicifrons*, *B.* sp., *Chamaedorea pauciflora*, *Desmoncus* sp., *Euterpe precatoria*, *Geonoma acaulis*, *G. camana*, *G.* cf. *jussieuana*, *G. piscicauda*,

Table 20. Species richness, density, and vertical distribution of palms in 0.5 ha of terra firme forest at Copal site, near Jenaro Herrera, lower Ucayali River valley, Peru. (Kahn and Mejia 1991)

	Height classes			
	0–1 m	1–10 m	>10 m	Total
Aiphanes ulei	2	5	0	7
Astrocaryum chambira	14	21	2	37
Astrocaryum cf. *chonta*	31	66	0	97
Bactris acanthospatha	3	10	0	13
Bactris bifida	7	13	0	20
Bactris piranga	2	2	0	4
Bactris sp. 3	8	17	0	25
Bactris sp. 4	1	7	0	8
Chamaedorea pauciflora	105	103	0	208
Chelyocarpus repens	7	14	0	21
Desmoncus cf. *leptospadix*	4	24	0	28
Euterpe precatoria	70	21	0	91
Geonoma acaulis	46	53	0	99
Geonoma camana	21	22	0	43
Geonoma oligoclona	33	9	0	42
Geonoma piscicauda	94	65	0	159
Geonoma poeppigiana	32	42	0	74
Geonoma pycnostachys	30	47	0	77
Geonoma spixiana	32	52	0	84
Geonoma cf. *tamandua*	4	9	0	13
Hyospathe elegans	415	221	0	636
Iriartea deltoidea	919	9	19	947
Iriartella stenocarpa	28	17	0	45
Jessenia bataua[a]	260	161	4	425
Lepidocaryum tessmannii	43	240	0	283
Maximiliana maripa	34	27	0	61
Oenocarpus mapora	0	5	0	5
Orbignya polysticha	0	1	0	1
Pholidostachys synanthera	31	51	0	82
Phytelephas macrocarpa	0	18	0	18
Scheelea bassleriana	18	13	0	31
Socratea exorrhiza	14	8	0	22
Socratea salazarii	81	36	0	117
Wettinia augusta	7	8	0	15
Total	2396	1417	25	3838

[a] *Jessenia bataua* subsp. *bataua*.

G. poeppigiana, *G. pycnostachys*, *G. spixiana*, *Hyospathe elegans*, *Jessenia bataua*, *Maximiliana maripa*, *Oenocarpus mapora*, *Prestoea asplundii*, *Socratea exorrhiza*, *Wettinia maynensis*.

In Ecuadorean Amazonia. The study site at Añangu is located in the northwest corner of Yasuní National Park (00°32′S; 76°26′W). The climate

Table 21. Palms with DBH > 10 cm

Species	Relative density[a]
A) Añangu, Ecuador. (Balslev et al. 1987)	
Astrocaryum chambira	0.87
Astrocaryum urostachys	0.62
Euterpe precatoria	0.12
Iriartea deltoidea	13.42
Jessenia bataua[b]	0.25
Maximiliana maripa	0.25
Socratea exorrhiza	0.37
Wettinia maynensis	0.75
B) Bolivian Amazonia. (Boom 1986)	
Astrocaryum aculeatum	2.31
Euterpe precatoria	8.17
Jessenia bataua[b]	0.92
Maximiliana maripa	0.15
Oenocarpus mapora	0.77
Scheelea princeps	0.15
Socratea exorrhiza	7.09

[a] Relative density = 100 × total number of individuals of species/total number of individuals in the sample.
[b] *Jessenia bataua* subsp. *bataua*.

is humid tropical with an average annual rainfall of 2500 mm and average annual temperature of 25.2 °C. Monthly rainfall is least from December to February. The transect (4020 m in two parallel lines) is "situated in unflooded forest mostly along a ridge of low rolling hills between 285 and 385 meters above sea level, dissected by numerous gullies" (Balslev et al. 1987).

The sample included 53 families and 244 species. Families with higher relative diversities (number of species in a family/total number of species ×100) are: Moraceae (9.0), Lauraceae (8.2), Meliaceae, Rubiaceae (6.2), Burseraceae (6.2), Caesalpiniaceae, Euphorbiaceae (4.1), Mimosaceae, Palmae, Myristicaceae (3.3), Annonaceae, Melastomataceae, Sterculiaceae (2.9). Total density was 728 trees per hectare (DBH > 10 cm), 14.6% of which was represented by three palm species, *Iriartea deltoidea*, *Socratea exorrhiza* and *Wettinia maynensis* (Table 21A). Palms in the understory were not inventoried, but the authors report an abundance of *Chamaedorea pinnatifrons*.

In Bolivian Amazonia. Boom's study site (1986) is located in the Bolivian Department of Beni, Province of Vaca Diez (11°45′S; 66°02′W). Average annual precipitation is about 1500 mm, and average annual temperature is 26.8 °C. All trees with DBH > 10 cm were inventoried in a 1-ha linear transect (10 × 1000 m). The five most important families in the forest surveyed are: Moraceae, Myristicaceae, Palmae, Leguminosae sensu lato, and Melastomataceae. Seven arborescent or medium-sized palm species were encountered (Table 21B). *Euterpe precatoria*, *Socratea exorrhiza*, and *Astrocaryum aculeatum* are the dominant species.

3.3.1.4 Conclusion: Palms in the Terra Firme Forests

Palm Life Forms and Structure of the Community. Most palm species found in the terra firme forests are small or medium-sized, understory species. In French Guiana, Sist (1989a) reported nine and ten understory species out of a total of 14 and 13 species found in 0.5 and 0.25 ha, respectively; de Granville (1989) found eight and six understory species of a total of ten and nine species in 1.26 ha and 0.18 ha surveyed, respectively. In Surinam, Van der Steege (1983) counted five understory species of the nine species found on 1.5 ha. In the lower Tocantins River, 8 of the 12 species found on 3.84 ha were understory species. In central Amazonia, of the 26 species found in 0.72 ha, 23 were understory palms, and only three were arborescent species. In the Peruvian terra firme forests, of 29 species found in 0.71 ha, 23 were understory species and six were arborescent species; and of 34 species found in 0.5 ha, 27 were understory species and seven were arborescent species.

In the understory, the relative importance of life forms is related to the abundance of some species (Table 22). For instance, the high percentages of multistemmed palms at Ferrocaño site, in Tucuruí forest, and in Galbao transect (84, 72.3, and 52% of the palm community, respectively) correspond to the high density of *Lepidocaryum tessmannii* in the first area, of *Astrocaryum gynacanthum*, *Bactris elegans*, and *Geonoma deversa* in the second, and of *Bactris gastoniana* in the third. The percentage of understory, single-stemmed palms is high in both plots at Piste de Saint Elie, with 34.2 and 32.9%, as a result of the abundance of *Astrocaryum sciophilum*, and low, 0.8 to 16.9%, in the other inventories where this life form is represented by *Chelyocarpus repens*, *Syagrus inajai*, *Geonoma poeppigiana*, *Geonoma triglochin* and *Pholidostachys synanthera*. Climbing (*Desmoncus* sp.) and subacaulescent palms with small leaves (*Aiphanes ulei*, *Geonoma acaulis*) were found only at the western sites. Subterranean-stemmed palms with large leaves are abundant in the forest of the lower Waki River valley (67.9%), represented by *Astrocaryum paramaca*, and in the central Amazonian forest (35.2%), represented by *Astrocaryum sociale*, *Orbignya polysticha*, *Scheelea* spec. nov.; they are less abundant or absent in the other forests inventoried.

Arborescent, single-stemmed species were largely or exclusively represented by seedlings and juveniles of the following species: *Iriartea deltoidea*, *Oenocarpus bacaba*, *Socratea exorrhiza*, *Euterpe precatoria*,

Table 22. Life forms of palms in terra firme forests on clayey soils

	% of the palm community (density)							
	LU		LRN	LT	GM	LW	P1	P2
Area surveyed (ha)	0.71	0.5	0.72	3.84	1.26	0.18	0.5	0.25
Tall palms								
No. of species	6	7	3	4	2	3	5	3
Single-stemmed (%)	5.7	42.1	31.2	25.2	24.9	8.0	32.4	47.9
Multistemmed (%)[a]	–	0.0	–	0.1	–	0.8	0.8	1.2
Small and medium-sized palms								
No. of species	23	27	23	8	8	6	9	10
Single-stemmed (%)	6.5	13.7	3.9	2.3	16.9	0.8	34.2	32.9
Multistemmed (%)	84	40.4	29.8	72.3	52.0	22.5	29.4	13.7
Acaulescent (%)	1.0	3.1	0.0	–	–	–	–	–
Subterranean-stemmed (%)	2.7	0.0	35.2	–	6.3	67.9	3.2	4.4
Climbing (%)	0.0	0.7	–	–	–	–	–	–

[a] Only seedlings and juveniles. Western Amazonia – LU: lower Ucayali River valley, near Jenaro Herrera. Central Amazonia – LRN: lower Rio Negro valley, near Manaus. Eastern Amazonia – LT: lower Tocantins River valley, near Tucuruí; GM: Galbao mountains; LW: lower Waki River valley; P1, P2: Piste de Saint Elie. (Kahn et al. 1988; Sist 1989a; Kahn and Mejia 1991).

Table 23. Density and vertical distribution of palms in six terra firme forests

Area surveyed (ha)	Height	
	<10 m	>10 m
Galbao mounts (1.26)	863 (99.9%)	1 (0.1%)
Lower Waki (0.18)	373 (98.9%)	4 (1.1%)
Lower Tocantins (3.84)	3442 (99.9%)	4 (0.1%)
Lower Rio Negro (0.72)	2322 (99.8%)	4 (0.2%)
Lower Ucayali (0.71)	6999 (99.9%)	5 (0.1%)
Lower Ucayali (0.50)	3813 (99.3%)	25 (0.7%)

Jessenia bataua, and *Maximiliana maripa*. Arborescent, multistemmed palms were encountered only at the eastern site, limited to a few seedlings and juveniles of *Euterpe oleracea*. All inventories point out the very low density of adult, arborescent palms: more than 98.9% of the palm communities are under 10 m in height (Table 23).

Palm Diversity in Terra Firme Forest Throughout the Amazon Basin. Palm species richness in terra firme forests is clearly higher in the central and western regions than in the eastern region of Amazonia, as has been pointed

Table 24. Palm species richness in terra firme forests on clayey soils

	No. of species per genus								
	LU		LRN	LT	GM	LW	P1	P2	S
Area surveyed (ha)	0.71	0.5	0.72	3.84	1.26	0.18	0.5	0.25	1.5
Aiphanes	0	1	0	0	0	0	0	0	0
Astrocaryum	1	2	2	1	1	1	2	2	2
Bactris	10	4	11	4	3	4	6	5	2
Chamaedorea	0	2	0	0	0	0	0	0	0
Chelyocarpus	1	1	0	0	0	0	0	0	0
Desmoncus	1	1	0	0	0	0	0	0	0
Euterpe	1	1	1	1	0	1	1	1	1
Geonoma	5	8	6	2	3	0	0	2	1
Hyospathe	1	1	0	0	1	0	0	0	0
Iriartea	0	1	0	0	0	0	0	0	0
Iriartella	1	1	1	0	0	0	0	0	0
Jessenia	1	1	0	0	0	0	1	1	0
Lepidocaryum	1	1	0	0	0	0	0	0	0
Maximiliana	1	1	0	1	0	0	1	0	1
Oenocarpus	1	1	2	1	1	1	1	1	1
Orbignya	1	1	0	0	0	0	0	0	0
Pholidostachys	1	1	0	0	0	0	0	0	0
Phytelephas	0	1	0	0	0	0	0	0	0
Scheelea	0	1	1	0	0	0	1	1	0
Socratea	1	2	1	1	1	1	1	0	1
Syagrus	0	0	1	1	0	1	0	0	0
Wettinia	1	1	0	0	0	0	0	0	0
Total: Species	29	34	26	12	10	9	14	13	9
Genera	16	21	9	8	6	6	8	7	7

Western Amazonia – LU: lower Ucayali River valley, near Jenaro Herrera. Central Amazonia – LRN: lower Rio Negro valley, near Manaus. Eastern Amazonia – LT: lower Tocantins River valley, near Tucuruí; GM: Galbao mountains; LW: lower Waki River valley; P1, P2: Piste de Saint Elie; S: Surinam. (Van der Steege 1983; Kahn et al. 1988; Sist 1989a).

out by Kahn et al. (1988). Several inventories from French Guiana and Surinam confirm the lower diversity of palms in eastern Amazonia (Table 24). In a forest of French Guiana, the species richness does not change when size of survey increases beyond 0.2 ha (Sist 1989a). In a western forest in the lower Ucayali River valley, 26 species were found in 0.08 ha, 32 in 0.3 ha, and 34 in 0.5 ha; the two surveys of 0.71 and 0.5 ha, 15 km apart but both located in the same physiographical unit, had in common only 17 of the 46 species inventoried.

Diversity of genera is remarkably higher in the western region. Four of the 21 genera found in 0.5 ha surveyed in the forest of the lower Ucayali River valley, *Aiphanes*, *Chamaedorea*, *Iriartea*, and *Wettinia*, and the species *Socratea salazarii*, are, in fact, more frequent in Andean piedmont

Table 25. Species richness, density, and vertical distribution of palms in 0.2 ha of a low forest (chamizal) on dry, white sands (orthic podzol), near Jenaro Herrera, lower Ucayali River valley, Peru

	Height classes			
	0–1 m	1–10 m	>10 m	Total
Bactris acanthospatha	1	2	0	3
Bactris simplicifrons	128	93	0	221
Geonoma acaulis	1	0	0	1
Iriartella stenocarpa	5	7	0	12
Jessenia bataua[a]	23	40	2	65
Mauritiella peruviana	5	7	3	15
Total	163	149	5	317

[a] *Jessenia bataua* subsp. *bataua*.

forests than in the low part of Peruvian Amazonia. Their occurrence here points to the influence of the Subandean palm flora on the western lowlands of the Amazon basin. The presence in this same forest at an elevation of 160 m above sea level of another genus, *Dictyocaryum*, known only to occur at elevations above 1500 m (Gentry 1986b), lends further evidence to the hypothesis of Subandean influence.

3.3.2 Forests on Sandy Soils

Podzols are characterized by an upper layer of white sand, very poor in nutrients, with organic matter only in its upper part. This sandy layer, up to several meters deep, lies on a hardpan, i.e., an indurated layer of organic matter and iron, 10 to 30 cm thick. Orthic podzols are dry soils; gleyic podzols are irregularly waterlogged soils. Floristic composition and architecture of the vegetation are different in each case. Trees are low-lying to tall in relation to water dynamics. These ecosystems are called "chamizal" in Peru and "campina" or "campinarana" in Brazil. The "campinarana", pseudo-campina, "is, in fact, neither a campina nor a true, dense forest . . ." (Guillaumet 1987, p. 247); the relative humidity ranges from 81 to 90% in a campina and from 91–97% in a campinarana (Ribeiro and Santos 1975). In Venezuelan, the terms "low Amazon caatinga", or "bana", and "tall Amazon caatinga" are used; both are on podzols with a permanent elevated groundwater table (Klinge et al. 1977; Klinge 1978; Klinge and Medina 1979; Klinge and Herrera 1983). In Guyana, woody vegetation on white-sandy soils (Berbice formation) is called "wallaba forest" (Davis and Richards 1933). Such soils also occur in Surinam (Zanderij series).

Fig. 40. *Jessenia bataua* subsp. *bataua*

On dry sandy soils (orthic podzol), most species show a clear tendency towards xeromorphy. Species richness of trees is low (Anderson et al. 1975; Prance and Schubart 1978; Anderson 1981), while epiphytic and nonvascular plants are diversified: 31 species in 17 genera of Orchidaceae (Braga 1977), and 34 bryophytes with five families of mosses and seven of liverworts (Lisbôa 1976). Lindeman and Moolenaar (1959) noted the low frequency of palms, mainly represented by *Astrocaryum paramaca*, *Oenocarpus bacaba*, and *Bactris* spp. on dry, sandy soils of Surinam. *Bactris arenaria* Barbosa Rodrigues, *B. balanophora* Spruce, and *B. campestris* are commonly found on these soils in both central northern and eastern Amazonia. Though just six palm species in five genera were inventoried on a 0.2-ha plot surveyed in a "chamizal" located in the lower Ucayali River basin (Table 25), palm density was significant (159 palms/0.1 ha). *Mauritiella peruviana* (Beccari) Burret is the most characteristic palm species of this ecosystem.

On a temporarily waterlogged gleyic podzol, vegetation is a forest with trees up to 35 m in height. Palm diversity is clearly higher, 23 species in 12 genera were found on a 0.27-ha plot surveyed. Palm density is very high (873/0.1 ha) with two dominant species (Figs. 39, 40): *Lepidocaryum tessmannii* (43.9%) in the understory, and *Jessenia bataua* subsp. *bataua* (36.0%), with 28 individuals, in the canopy. Though also occurring in uplands on this waterlogged gleyic podzol, many species are more com-

Table 26. Species richness and density of palms in 0.27 ha of a forest on waterlogged, white sands (gleyic podzol), near Jenaro Herrera, lower Ucayali River valley, Peru

	Height classes			
	0–1 m	1–10 m	>10 m	Total
Astrocaryum cf. *chonta*	2	2	0	4
Bactris mitis	2	3	0	5
Bactris piranga	0	1	0	1
Bactris simplicifrons	2	2	0	4
Bactris sphaerocarpa	4	6	0	10
Bactris sp. 1	9	10	0	19
Bactris sp. 2	0	4	0	4
Euterpe precatoria	17	7	0	24
Geonoma acaulis	51	79	0	130
Geonoma camana	0	1	0	1
Geonoma oligoclona	0	1	0	1
Geonoma piscicauda	2	0	0	2
Geonoma poeppigiana	6	11	0	17
Geonoma spixiana	1	2	0	3
Hyospathe elegans	14	9	0	23
Iriartella stenocarpa	88	27	0	115
Jessenia bataua[a]	565	255	28	848
Lepidocaryum tessmannii	119	915	0	1034
Mauritia flexuosa	36	18	0	54
Orbignya polysticha	0	2	0	2
Pholidostachys synanthera	13	30	0	43
Socratea exorrhiza	5	0	0	5
Socratea salazarii	0	8	0	8
Total	936	1393	28	2357

[a] *Jessenia bataua* subsp. *bataua*.

monly found in seasonal swamp forests: *Astrocaryum* cf. *chonta*, *Euterpe precatoria*, *Geonoma acaulis*, *Jessenia bataua* subsp. *bataua*, *Mauritia flexuosa*, and *Socratea exorrhiza* (Table 26).

An association of three species, *E. catinga*, *Jessenia bataua* subsp. *bataua*, and *Mauritia carana*, observed near Manaus as well as near Iquitos, is characteristic on sandy, poorly drained soils.

3.4 Flooded Forests

Flooded forests are located in the floodplains of the major rivers, and also in the network of valleys which is scoured through uplands by small streams collecting water from slopes and plateaus. Classification and terminology of Amazonian wetland forest types have been proposed and discussed by many authors (Ule 1908; Fanshawe 1952; Ducke and Black 1953; Aubréville 1961; Takeuchi 1962; Irmler 1977; Prance 1979; de Granville 1986, 1988). Klinge et al. (1990) summed up the general characteristics of these ecosystems in South America. The following classification does not pretend to replace the former; it simply allows a coherent presentation of flooded-forest palms.

In upland valleys, the seasonal swamp forests stand on soils which are permanently waterlogged and irregularly flooded by streams. Inundations occur during the rainy season immediately following heavy rainfall. Floodplain forests are inundated by the river. Timing and frequency of flooding, nutrient content of water and sediments, and water acidity all contribute to differentiate: forests on alluvial soils periodically flooded by whitewater rivers, i.e., with a coffee and milk color which is due to a heavy sediment load; forests periodically flooded by blackwater rivers, i.e., light coffee-colored waters; and swamp forests permanently flooded by black or clear rivers, these with a light or temporary sediment load. Flooded forests which receive tidal influence are considered as part of the forests on periodically flooded alluvial soils or of the permanent swamp forests because of the similarity of their palm communities. Mangrove forests are very poor in palms. *Bactris major* is the only species known to grow under these extreme ecological conditions.

3.4.1 Seasonal Swamp Forests

These forests are located in the bottoms of the valleys which dissect the uplands collecting the water from slopes and plateaus. Soils are permanently waterlogged, generally characterized by a gley; inundations occur irregularly during the rainy season. Richards (1969) concluded that forests on swampy soils are always poorer in species than those on well-drained soils. The inventory of Porto et al. (1976) includes 95 dicotyledonous tree species in 33 families. Among the largest trees, Takeuchi (1960) noted *Scleronoma micranthum* (Bombacaceae), *Vitex sprucei* (Verbenaceae), *Chromolucama rubriflora* (Sapotaceae), and listed 176 palms from a 2200-m^2 plot. Monocotyledons are often abundant in patches of single-species stands (Heliconiaceae, Marantaceae, Rapateaceae, Strelitziaceae, Zingiberaceae). Terrestrial and hemiepiphytic Cyclanthaceae are also frequent. According to de Granville (1978), monocotyledons account for 12 to 31% of the species of the flowering plants in Guianan seasonal swamp forests. Ferns are also abundant in the understory of this ecosystem (de Granville 1984). Palms dominated the canopy as well as the understory of the seasonal swamp forests.

Examples from Eastern Amazonia. In French Guiana at Piste de Saint Elie site (Fig. 30a), Sist (1989a) reported 11 species and 6 genera in 0.24 ha of seasonal swamp forest on waterlogged soils. The dominant species are

Table 27. Species richness and density of palms in 0.24 ha of a seasonal swamp forest at Piste de Saint Elie, French Guiana. (Sist 1989a)

	Seedlings	Juveniles and adults	Total
Astrocaryum paramaca	45	59	104
Astrocaryum sciophilum	6	0	6
Bactris aubletiana	5	0	5
Bactris constanciae	1	1	2
Bactris oligocarpa	1	0	1
Bactris raphidacantha	1	0	1
Bactris simplicifrons	2	0	2
Euterpe oleracea	4	1	5
Geonoma oldemanii	74	190	264
Jessenia bataua[a]	261	138	399
Scheelea sp.	4	8	12
Total	404	397	801

[a] *Jessenia bataua* subsp. *oligocarpa*.

Jessenia bataua subsp. *oligocarpa*, *Geonoma oldemanii* and *Astrocaryum paramaca*. (Table 27). The palm most commonly found on waterlogged soils, *Euterpe oleracea* is not well represented in this coastal forest, though it forms very dense populations on waterlogged soils elsewhere (Oldeman 1969), usually associated with the dicotyledons *Virola surinamensis* (Myristicaceae) and *Symphonia globulifera* (Clusiaceae), and the palms *Socratea exorrhiza*, *Geonoma baculifera*, *G. oldemanii*, sometimes *Jessenia bataua* subsp. *oligocarpa* and *Hyospathe elegans* (de Granville 1978, 1990).

In the lower Tocantins River basin (Fig. 34a), six species and six genera were found in each of two 0.24-ha inventories near the villages of Chiquerão and Remansão (Table 28A,B). Of the seven species, four are arborescent, *Euterpe oleracea*, *Maximiliana maripa*, *Orbignya phalerata*, *Socratea exorrhiza*, and three are understory palms, *Astrocaryum gynacanthum*, *Bactris elegans*, and *Syagrus inajai*. *O. phalerata*, the babassu palm, was represented by only two adults in 0.24 ha at the Remansão site and was absent in the Chiquerão plot. This species becomes the most abundant arborescent palm southwards, as shown by an inventory carried out in the Serra dos Carajas (Scariot et al. 1989), where eight species and eight genera of palms were found in 1 ha surveyed (Table 29).

Examples from Central Amazonia. In a seasonal swamp forest on a sandy gleysol, North of Manaus, two 0.12-ha plots yielded six and seven species, respectively. Twenty palms over 10 m in height were counted in the aggregated 0.24-ha total (Table 30). Most of them reached the canopy and were up to 30 m in height. *Euterpe precatoria* and *Jessenia bataua* subsp. *bataua* were the dominant species and represented together 95.6% of the com-

Table 28. Species richness, density, and vertical distribution of palms in two seasonal swamp forests on waterlogged soils located in narrow upland valleys in the lower Tocantins River valley, Brazil (0.24 ha surveyed in each case)

	Height classes			
	0–1 m	1–10 m	>10 m	Total
A) Seasonal swamp forest near Chiquerão				
Astrocaryum gynacanthum	2	37	0	39
Bactris elegans	0	1	0	1
Euterpe oleracea	4	77	25	106
Geonoma baculifera	12	14	0	26
Maximiliana maripa	1	16	1	18
Socratea exorrhiza	6	17	9	32
Total	25	162	35	222
B) Seasonal swamp forest near Remansão				
Astrocaryum gynacanthum	4	3	0	7
Euterpe oleracea	31	70	13	114
Maximiliana maripa	0	1	1	2
Orbignya phalerata	485	63	2	550
Socratea exorrhiza	0	3	0	3
Syagrus inajai	0	2	0	2
Total	520	142	16	678

Table 29. Species richness and density of palms in 1 ha of a seasonal swamp forest, Serra dos Carajás, Brazil. (After Scariot et al. 1989)

	Seedlings	Juveniles	Adults	Total
Astrocaryum gynacanthum	78	80	38	196
Bactris sp.	4	36	1	41
Euterpe oleracea	334	9	58	401
Geonoma baculifera	84	396	222	702
Maximiliana maripa	0	2	0	2
Oenocarpus distichus	0	0	3	3
Orbignya phalerata	2177	199	17	2393
Socratea exorrhiza	117	105	15	237
Total	2794	827	354	3975

munity. Only three palm species in the understory were present: *Bactris trailiana*, *Manicaria martiana*, and *Orbignya spectabilis*.

Examples from Western Amazonia. In a seasonal swamp forest in an upland valley of the lower Ucayali River valley (Peru), 18 species were

Table 30. Species richness, density, and vertical distribution of palms in 0.24 ha of a seasonal swamp forest, near Manaus, lower Rio Negro valley, Brazil

	Height classes			
	0–1 m	1–10 m	>10 m	Total
Bactris trailiana	8	68	0	76
Euterpe precatoria	794	21	5	820
Jessenia bataua[a]	1092	485	15	1592
Manicaria martiana	7	1	0	8
Mauritiella aculeata	3	1	0	4
Orbignya spectabilis	0	19	0	19
Socratea exorrhiza	3	1	0	4
Total	1907	596	20	2523

[a] *Jessenia bataua* subsp. *bataua*.

found in 1 ha surveyed (Table 31). Seven of these were located in a ±500 m^2 area with dryer soils within the wetland parcel. These species, which are very common in uplands (Kahn and Mejia 1991), cannot be considered representative of this swamp community. Six species represent 89.2% of the palm community (over 1 m in height): *Bactris monticola* (36.4%), *Jessenia bataua* (15.2%), *Euterpe precatoria* (10.5%), *Mauritia flexuosa* (10.5%), *Geonoma acaulis* (10.0%), and *Socratea exorrhiza* (6.7%). Palm basal area represents 36.3% of the total basal area (26.3 m^2/ha for DBH > 15 cm), with 279 trunks, 172 of which reach more than 10 m in height. Their crowns constitute a major component of the forest canopy. *B. monticola* is the dominant species in the understory on the waterlogged soil while *Lepidocaryum tessmannii* reaches a very high density on the small "island" of dryer soil.

In a seasonal swamp forest in the upper Huallaga River valley (Peru), the fifteen species found were present in each of the two 0.8-ha plots surveyed (Table 32: density values for each species are lumped into 1.6 ha). Nine of them represent 93.7% of the community: *Geonoma brongniartii* (15.4%), *Jessenia bataua* (14.5%), *Euterpe precatoria* (11.9%), *Geonoma macrostachys* (10.9%), *Oenocarpus mapora* (10.7%), *Socratea exorrhiza* (9.9%), *Iriartea deltoidea* (9.4%), *Wettinia augusta* (6.3%), and *Mauritia flexuosa* (4.7%). Density of palms greater than 10 m in height is low (1.6% of the community) with 43 trunks, 22 of which are *M. flexuosa*. Structural data from the upper Huallaga valley are biased by the high degree of disturbance of the forest and cannot be considered representative. The density of trunked palms of the three arborescent species, *E. precatoria*, *J. bataua*, and *M. flexuosa*, is particularly low. Many palms had been cut down, and a lot of decomposing trunks (mainly of *J. bataua* and *M. flexuosa*, which are the larger species) lay on the forest floor. All three provide useful

Table 31. Species richness, density, and vertical distribution of palms in a 1-ha plot of a seasonal swamp forest, near Jenaro Herrera, lower Ucayali River valley, Peru. (Kahn and Mejia 1990)

	Height classes			
	0–1 m	1–10 m	>10 m	Total
Astrocaryum cf. *chonta*	–	13	0	13
Bactris acanthocarpoides[a]	–	3	0	3
Bactris monticola	–	866	0	866
Bactris simplicifrons	–	18	0	18
Bactris sp.[a]	–	5	0	5
Euterpe precatoria	–	205	45	250
Geonoma acaulis	–	237	0	237
Geonoma leptospadix[a]	–	1	0	1
Geonoma oligoclona	–	10	0	10
Geonoma piscicauda[a]	–	1	0	1
Geonoma spixiana[a]	–	4	0	4
Jessenia bataua[b]	–	313	48	361
Lepidocaryum tessmannii[a]	–	151	0	151
Mauritia flexuosa	–	218	32	250
Oenocarpus mapora	–	39	1	40
Orbignya polysticha[a]	–	9	0	9
Socratea exorrhiza	–	113	46	159
Wettinia augusta	–	2	0	2
Total		2208	172	2380

[a] These species are common in terra firme forests. They occur here on a $\pm 500\,m^2$ island with dryer soil.
[b] *Jessenia bataua* subsp. *bataua*.

products, such as palm heart, oil, edible fruit (see Chap. 7). They have been extensively cut down by the numerous new inhabitants who have settled in this valley, attracted by job prospects at a 4500-ha African oil palm plantation.

3.4.2 Forests on Alluvial Soils Periodically Flooded by Whitewater

These forests correspond to the last stage of the primary succession occuring on alluvial soils deposited by whitewater rivers. The older the forest, the shorter is the time of flooding, which varies from several weeks to 5 months. These lands are called “várzea” in Brazil (Prance 1979; Junk 1984) and “restinga” in Peru (Encarnación 1985; López Parodi and Freitas 1990). The forests which stand the farthest from the river bed are flooded only during years of highest flooding, not every year. In the eastern region of Brazilian Amazonia, such forests also receive the effect of tide. Prance (1979) defined them as tidal várzea.

Table 32. Species richness, density, and vertical distribution of the palms in 1.6 ha of a seasonal swamp forest, near Uchiza, upper Huallaga River valley, Peru. (Kahn and Mejia 1990)

	Height classes			
	0–1 m	1–10 m	>10 m	Total
Astrocaryum carnosum	10	22	0	32
Bactris maraja	2	15	0	17
Bactris sp.	32	50	0	82
Chamaedorea cf. *lanceolata*	0	2	0	2
Desmoncus sp.	0	7	0	7
Euterpe precatoria	239	88	1	328
Geonoma acaulis	10	23	0	33
Geonoma brongniartii	103	320	0	423
Geonoma macrostachys	152	147	0	299
Iriartea deltoidea	153	104	0	257
Jessenia bataua[a]	231	160	7	398
Mauritia flexuosa	22	85	22	129
Oenocarpus mapora	109	184	0	293
Socratea exorrhiza	206	57	9	272
Wettinia augusta	127	42	4	173
Total	1396	1306	43	2745

[a] *Jessenia bataua* subsp. *bataua*.

Data from Eastern Amazonia. According to Anderson (1988), the flooded forests of the Amazon estuary are characterized by a relatively low biological diversity and pronounced dominance by a few tree species. Among them are *Euterpe oleracea*, the populations of which cover an estimated area of 10 000 km^2 (Calzavara 1972), *Astrocaryum murumuru*, and *Raphia taedigera*. Annually flooded alluvial soils are colonized by *Astrocaryum jauari* and *Bactris maraja* in the lower Tocantins River valley.

Data from Central Amazonia. There are a few data on palms of flooded alluvial soils (várzea) from central Amazonia. *Astrocaryum* cf. *chonta* is a frequent species on these soils near Manaus; *A. jauari*, *Bactris concinna* Martius, *B. maraja* occur in a narrow strip along the river bed. *Euterpe oleracea* and *Oenocarpus mapora* are frequently cultivated, introduced from the eastern and western region, respectively.

Examples from Peruvian Amazonia. The canopy of these periodically flooded forests is broken, more open than that of the terra firme forest. Floristic composition of trees is dominated by Moraceae, mainly the genus *Ficus* (Lamotte 1990; Salo et al. 1986).

In a forest on alluvial soil (restinga) located in the lower Ucayali River valley, 11 species were found on the 0.4-ha plot (Table 33). Among them,

Table 33. Species richness, density, and vertical distribution of palms in 0.4 ha of a forest on periodically flooded alluvial soils, near Jenaro Herrera, lower Ucayali River valley, Peru. (Kahn and Mejia 1990)

	Height classes			
	0–1 m	1–10 m	>10 m	Total
Astrocaryum cf. *chonta*	722	328	0	1050
Bactris bifida	183	95	0	278
Bactris maraja	0	1	0	1
Bactris monticola	7	5	0	12
Desmoncus polyacanthos	0	1	0	1
Euterpe precatoria	6	2	0	8
Geonoma acaulis	78	34	0	112
Oenocarpus mapora	0	2	0	2
Phytelephas macrocarpa	25	841	0	866
Scheelea brachyclada	372	97	7	476
Socratea exorrhiza	0	6	0	6
Total	1393	1412	7	2812

Fig. 41. *Bactris bifida*

five represent 98.9% of the community: *Astrocaryum* cf. *chonta* (37.3%), *Phytelephas macrocarpa* (30.8%), *Scheelea brachyclada* (16.9%), *Bactris bifida* (9.9%; Fig. 41), and *Geonoma acaulis* (4.0%). Density of palms over 10 m in height is very low (0.3%), and very high for individuals between 1–10 m (353 palms/0.1 ha, 50.2% of the community). *Astrocaryum* cf.

Fig. 42. *Phytelephas macrocarpa*

chonta and *Phytelephas macrocarpa* (Fig. 42), a single and a multistemmed species, respectively, dominate the 1–10 m layer; both species rarely reach more than 10 m in height. Density of palms under 1 m in height is high (49.5% of the community) in spite of the 3–4 month annual flooding.

In a forest on periodically flooded alluvial soils located in the upper Huallaga valley, 15 species were found on 0.4 ha surveyed (Table 34). Four of them constitute 86.0% of the community: *Astrocaryum carnosum* (49.5%), *Euterpe precatoria* (20.8%), *Oenocarpus mapora* (11.1%), and *Geonoma macrostachys* (4.6%). *A. carnosum*, a medium-sized, multistemmed species, is the dominant palm in the understory. Density of palms over 10 m in height is low (0.7% of the community).

An inventory of all trees with DBH > 10 cm carried out on a 1-ha plot of Cosha Cashu in Manu National Park by Gentry (1985) lists 38 individuals of *Iriartea deltoidea* and 37 individuals of *Astrocaryum* cf. *chonta*. Some *Bactris* sp. and *Scheelea* sp. were also found in the area surveyed.

Examples from Ecuadorean Amazonia. At Añangu, a transect (2100 m consisting of seven parallel lines within an area of 340 × 120 m) was located in a flat floodplain at 285 m above sea level near the junction of Río Añangu and Río Napo (Balslev et al. 1987). The sample included 44 families and 149

Table 34. Species richness, density, and vertical distribution of palms in 0.4 ha of a forest on periodically flooded alluvial soils, near Uchiza, upper Huallaga River valley, Peru. (Kahn and Mejia 1990)

	Height classes			
	0–1 m	1–10 m	>10 m	Total
Astrocaryum carnosum	230	352	0	582
Bactris sp.	4	12	0	16
Chamaedorea cf. *lanceolata*	5	7	0	12
Chelyocarpus ulei	8	10	0	18
Euterpe precatoria	176	67	1	244
Geonoma acaulis	3	4	0	7
Geonoma macrostachys	35	19	0	54
Geonoma maxima	1	2	0	3
Iriartea deltoidea	0	8	0	8
Jessenia bataua[a]	1	2	1	4
Mauritia flexuosa	1	2	3	6
Oenocarpus mapora	37	93	0	130
Phytelephas macrocarpa	2	34	0	36
Scheelea sp.	9	11	0	20
Socratea exorrhiza	16	16	3	35
Total	528	639	8	1175

[a] *Jessenia bataua* subsp. *bataua*.

species. The families with highest relative diversity were: Moraceae (8.7), Rubiaceae (8.0), Family unknown (6.0), Lauraceae (4.7), Caesalpiniaceae, Meliaceae, Sterculiaceae (4.0), Palmae, Cecropiaceae, Chrysobalanaceae, Flacourtiaceae (3.4). The sample included 420 trees with DBH > 10 cm, or 417 trees per hectare, 4.3% of which were represented by *Iriartea deltoidea*. Only five palm species with DBH > 10 cm were found (Table 35). *Astrocaryum urostachys* and *Scheelea* sp. were the dominant palm species. The authors reported the abundance of *Geonoma brongniartii* in the understory.

3.4.3 Forests Periodically Flooded by Blackwater

These forests are flooded for 5–9 months according to their distance from the main river. Blackwater has a high content of humic and fluvic acids, and the sediment load is almost nil. Soils consist of clay, silt, and sand material (Adis 1984). These wetlands, named "igapó" in Brazil (Prance 1979) and "tahuampa" in Peru (Encarnación 1985; López Parodi and Freitas 1990), present the lowest species diversity: 60 species (DBH > 10 cm) on 1 ha surveyed in an igapó-forest near Belém, Brazil (Black et al. 1950); 180 species (DBH >30 cm) on a 5-ha plot (Pires 1978). Adis (1984) distin-

Table 35. Palms with DBH > 10 cm in a forest on alluvial soils at Añangu, Ecuadorean Amazonia. (Balslev et al. 1987)

Species	Relative density
Astrocaryum urostachys	10.7
Iriartea deltoidea	4.3
Mauritia flexuosa	1.2
Phytelephas macrocarpa	4.3
Scheelea sp.	9.5

guished lower and upper igapó-forests. The lower igapó directly borders river courses; the unflooded period is very short; species richness (19 species/5 ha) is very low (Revilla 1981); *Myrciaria dubia* (Myrtaceae) is a dominant species. The upper igapó is inundated for 5–6 months, the trees reach 35 m in height, and species richness is still very low (30 species/2.5 ha). In the Anavilhanas archipelago of the lower Rio Negro, an island on clayey soils inundated for 2 months annually was surveyed by Rodrigues (1961). He recorded 51 species in the igapó-forest, a lower species richness than is exhibited by upland forests of the same region (Prance et al. 1976).

There are a few data on palms in this ecosystem. Palm species richness is very low. The most common species are *Astrocaryum jauari*, an arborescent, multistemmed palm which can reach up to 20 m in height, *Bactris concinna*, and *B. maraja*, both small, multistemmed palms, up to 10 m in height. These clustered species constitute dense, monospecific patches in small areas. In Ecuadorean Amazonia in Laguna Cuyabeno, a blackwater lake, Borgtoft Pederson and Balslev (1990) reported 2640 seedlings, 82 acaulescent juveniles, and 27 trunked individuals of *Astrocaryum jauari* in 0.1 ha. *Leopoldinia pulchra* forms dense populations in the igapó-forests near Manaus.

3.4.4 Permanently Flooded Swamp Forests

In eastern Amazonia, coastal swamps are under tidal influence. *Mauritia flexuosa* forms dense stands in Orinoco and Amazon estuaries (Ducke and Black 1953; Pires and Prance 1985; Klinge et al. 1990), and along coastal Guianan creeks. This species is found in association with *Montrichardia* (Araceae) and *Euterpe oleracea* colonizing mud and sand banks (Richards 1952; Strudwick and Sobel 1988). *Elaeis oleifera* and *Bactris penicillata* occur in Guianan river estuaries; both species are known only from restricted areas in French Guiana, located in the lower Mana River and in Kaw region, respectively. Permanent swamp forests in the subcoastal plain on marine sediments consist of almost pure stands of *Euterpe oleracea* (Fig. 43);

Fig. 43. *Euterpe oleracea* swamp

this formation is called "pinotière" in French Guiana, from "pinot", the local name of this species (Oldeman 1969). In central and western Amazonia, the permanent swamp vegetation is heavily dominated by *Mauritia flexuosa* (Fig. 44). These swamps usually stand in the narrow depressions which lie parallel to river beds. The substratum is clayey and retains rainfall which blends with the river water only at the time of highest flooding (Spruce 1871). *Mauritia* swamps are called "buritizais" in Brazil, "aguajales" in Peru, "cananguchales" in Colombia, "achuales" in Ecuador, and "morichales" in Venezuela and in some parts of Colombia, from the vernacular names of the species, "burití", "aguaje", "canangucha", "acho", and "moriche", respectively. Density of dicotyledons is low, with some large trees, however, such as *Hura crepitans* (Euphorbiaceae).

Soils of both *E. oleracea* and *M. flexuosa* swamps are characterized by an accumulation of slightly decomposed organic matter, several meters in depth, in acidic water. This organic matter is chiefly composed of dead leaves, inflorescences, and infructescences of the palms. The close relationship of *M. flexuosa* to water is suggested by Spruce's comments (1871, p. 78): "The prevalent opinion, or rather superstition, throughout Amazonia and Guyana is that the Mauritia has the power of attracting water to itself wherever planted." The streams which flow from *Mauritia* formations are blackwater (Klinge et al. 1990). Spruce (1871, p. 78) also noted: "... at the head of each of the "caños", or rivulets, that run into the Upper Rio Negro and Orinoco, there is a swamp where the predominant vegetation is *Mauritia flexuosa* ...".

Fig. 44. *Mauritia flexuosa* swamp

In a permanently flooded swamp forest (aguajal) in the lower Ucayali River basin, 11 species were found in 1 ha surveyed (Table 36). Among them, four represent 90.2% of the palm community (greater than 1 m in height): *Mauritia flexuosa* (54.5%), *Geonoma acaulis* (21.3%), *Oenocarpus mapora* (10.2%), and *Euterpe precatoria* (4.2%). The canopy of this swamp forest is essentially composed of *M. flexuosa* crowns. Density of palms over 10 m in height is high, with 207 individuals per hectare. Palm basal area represents 55% of the total basal area (31.1 m^2/ha), with 336 trunks, including 230 individuals of *M. flexuosa* (138 adults, and 92 trunked juveniles). Palm density in the understory (1–10 m) is 98 palms/0.1 ha, half of which are acaulescent and trunked juveniles of *M. flexuosa*. The density of *Astrocaryum* cf. *chonta* and *Scheelea brachyclada* is very low, though both are dominant species on the two "restingas" which bound the *M. flexuosa* swamp depression. *Phytelephas macrocarpa*, also abundant in the neighboring "restingas", is absent in the swampy depression.

In the eastern Andean piedmont of Peru, a density per hectare of 351 trunked palms (CV: 26%) and 297 acaulescent juveniles (CV: 43%) was estimated by Gonzáles Rivadeneyra (1971) from 20, 0.05-ha plots surveyed. Near Iquitos, along the Itaya River, a density per hectare of 246 trunked palms was estimated by Salazar and Roessl (1977) from ten 0.5-ha plots

Table 36. Species richness, density, and vertical distribution of palms in 1 ha of a permanently flooded *Mauritia* swamp, near Jenaro Herrera, lower Ucayali River valley, Peru. (Kahn and Mejia 1990)

	Height classes			
	0–1 m	1–10 m	>10 m	Total
Astrocaryum cf. *chonta*	–	5	0	5
Bactris concinna	–	15	0	15
Bactris maraja	–	42	0	42
Bactris monticola	–	26	0	26
Desmoncus polyacanthos	–	20	0	20
Euterpe precatoria	–	16	34	50
Geonoma acaulis	–	252	0	252
Mauritia flexuosa	–	478	167	645
Oenocarpus mapora	–	119	2	121
Scheelea brachyclada	–	1	0	1
Socratea exorrhiza	–	3	4	7
Total		977	207	1184

(CV: 24.7%). Similar densities of *Mauritia flexuosa* were reported by Urrego Giraldo (1987) from Colombian Amazonia.

3.4.5 Conclusions: Palms in Flooded Forests

Floristic Composition of Palm Communities. The occurrence of many palm species in several inventories carried out in flooded forests indicates a rather high floristic similarity between these ecosystems. In the lower Ucayali River valley, the six species which are common to the three inventories represent 26.6, 38.9, and 65.8% of the palm communities (over 1 m in height) in the forest on periodically alluvial soils, in the *M. flexuosa* swamp, and in the seasonal swamp forest, respectively. These two latter forests have seven species in common, which constitute 93.4 and 76.3% of their palm communities, respectively. In the upper Huallaga River valley, 11 of the 18 species are common to both ecosystems, and represent 77.4 and 93.4% of the communities in the seasonal swamp forest and in the forest on periodically flooded alluvial soils, respectively. Palm community characteristics of each of the wetland forest ecosystems are, nevertheless, very distinct, and associations of palm species can be identified (Kahn and Mejia 1990):

1. In eastern Amazonia, seasonal swamp forests are widely dominated by *Euterpe oleracea* (Oldeman 1969; Kahn 1986a; de Granville 1990) and by *Jessenia bataua* subsp. *oligocarpa* in the Guianas (de Granville 1978; Sist and Puig 1987). In central and western Amazonia, this forest ecosystem is

dominated by four species, *Euterpe precatoria*, *Jessenia bataua* subsp. *bataua*, *Mauritia flexuosa*, and *Socratea exorrhiza*, which together constitute an important component of the canopy. *J. bataua* subsp. *bataua* is a very characteristic palm of seasonal swamp forests on waterlogged soils; this species, which also forms dense populations on waterlogged sandy soils, was rarely found in other flooded ecosystems (see Sects. 3.3.2 and 4.2.2). Five other species are particularly common in this ecosystem in the Andean piedmont: *Geonoma brongniartii*, *G. macrostachys*, *Iriartea deltoidea*, *Oenocarpus mapora*, and *Wettinia augusta*. Some species are patchily distributed, such as *Asterogyne guianensis*, *Geonoma oldemanii*, and *Manicaria saccifera* in the Guianas, *M. martiana* in central Amazonia, *M.* sp. in Peru, *Geonoma baculifera* in the eastern region, *Iriartea deltoidea* in western Amazonia (in both seasonal swamp and terra firme forests), *Elaeis oleifera* and *Mauritiella aculeata* throughout the basin. Other species are widely distributed, such as *Geonoma acaulis* and *Oenocarpus mapora* in the western region, and *Socratea exorrhiza* throughout the basin.

2. In eastern Amazonia, forests on periodically flooded alluvial soils are dominated by *Euterpe oleracea* and *Astrocaryum murumuru*. In central Amazonia, *Astrocaryum* cf. *chonta* is common. In western Amazonia, these forests are characterized by the association of several species of three genera: *Astrocaryum* (*A. carnosum*, *A.* cf. *chonta*, or *A. urostachys*, three closely related species), *Phytelephas* (*P. macrocarpa*), and *Scheelea* spp. In contrast to other flooded forests, arborescent palms are not dominant in the canopy of these forests. On the other hand, palms are the major component of the understory because of the abundance of *Astrocaryum* spp. and/or *Phytelephas macrocarpa*. Dense populations of the latter are known as "yarinales" and are usually said to indicate good soils. The association of these three palm genera also occurs in an Ecuadorean floodplain forest on alluvial soils (Balslev et al. 1987).

3. Forests which are periodically flooded by blackwater streams are characterized by *Leopoldinia* spp. in central Amazonia and along the Rio Negro River, and by riparian palms such as *Astrocaryum jauari* and *Bactris maraja*, the fruit of which is disseminated by fishes (Gottsberger 1978; Goulding 1980). These species also occur on the banks of most Amazonian rivers throughout the basin (Spruce 1871; Barbosa Rodrigues 1903; Huber 1906). *Desmoncus polyacanthos*, a climbing palm, is commonly found in these forests, as well as in most wetlands, in low vegetation, and in open forests.

4. Swamp forests in permanently flooded depressions are mainly composed of *Mauritia flexuosa* in central and western Amazonia, and *Euterpe oleracea* in eastern Amazonia, both species reaching very high densities. Juvenile plants dominate the understory and adult plants dominate the canopy. *Socratea exorrhiza* is the only other arborescent palm which is frequent in these swamps. In the coastal Atlantic strip, swamps under tidal influence are dominated by the association *E. oleracea* and *M. flexuosa*.

Table 37. Species richness of palms in flooded forests of Amazonia

	Species	Genera	Area surveyed (ha)
Forests periodically flooded by blackwater:			
Lower Ucayali[a]	2–5	2	1
Forests on periodically flooded alluvial soils:			
Lower Ucayali[a]	11	9	0.4
Upper Huallaga[a]	15	13	0.4
Permanently flooded swamp forests:			
Lower Ucayali[a]	11	9	1
Seasonal swamp forests:			
French Guiana[b]	11	6	0.24
Lower Tocantins, Brazil[c]	6	6	0.12
Lower Tocantins, Brazil[c]	6	6	0.12
Serra dos Carajas, Brazil[d]	8	8	1
Lower Rio Negro[e]	6–7	6–7	0.12
Lower Ucayali[a]	18	11	1
Upper Huallaga[a]	15	12	0.8
Upper Huallaga[a]	15	12	0.8

[a] Kahn and Mejia 1990.
[b] Sist 1989a.
[c] Unpublished data.
[d] Scariot et al. 1989.
[e] Kahn and Castro 1985.

Palm Diversity. Palm inventories were systematically carried out in all flooded forest types in Peruvian Amazonia, allowing the comparison of palm diversity in those ecosystems. A few species (two to five species per ha) occur in the forest periodically flooded by blackwater river. Palm species richness in each of the other wetland forests in the lower Ucayali River valley is not significantly different: 11 species-9 genera in 0.4 ha on periodically flooded alluvial soils; 11 species-9 genera in 1 ha in a *M. flexuosa* swamp; 11 species-9 genera in 1 ha in the seasonal swamp forest (without counting seven species and two genera, which are restricted to a small area on dryer soil patches). Palm diversity is slightly higher in wetland forests in the upper Huallaga River valley: 15 species-13 genera in 0.4 ha on periodically flooded alluvial soils; 15 species-12 genera in each of both 0.8-ha plots surveyed in a seasonal swamp forest. Palm diversity in seasonal swamp forests in central and eastern Amazonia is also low (Table 37).

Structure of Palm Communities. In the swamp forests, density of palms over 10 m in height is very high: 172 and 207 individuals per ha, respectively, in a seasonal and permanent swamp forest in Peruvian Amazonia. This is not so with forests on periodically flooded alluvial soils; just 18 trunked palms over 10 m in height were found in 1 ha on alluvial soils in Peruvian Amazonia.

3.5 Montane Forests

Most Amazonian palms do not occur in the wet forests at high elevation. Nevertheless, *Astrocaryum huicungo*, *Iriartea deltoidea*, *Jessenia bataua* subsp. *bataua*, *Oenocarpus mapora*, *Phytelephas macrocarpa*, which are all frequent in wetland forests at low elevation, are the most common species between 600 m and 1000 m in the western region. In French Guiana, *O. bacaba*, *Bactris simplicifrons*, *Hyospathe elegans*, all frequent in terra firme forests, are found on the highest forested tops which culminate at 800 m. Other species are endemic to montane or submontane forests where they can grow at medium elevation. These include *Geonoma euspatha* in the Guianas, Venezuela, and Colombia, and *Geonoma triglochin* also found in Amapa and Peru (Wessels Boer 1968). With increasing altitude the following palms become more frequent: *Aiphanes* spp., *Chamaedorea* spp., some *Geonoma* species such as *G. helminthoclada*, which reach their greatest size at these altitudes, *Prestoea asplundii*, and the tall *Iriartea deltoidea*. Arborescent *Dictyocaryum lamarckianum* and *Prestoea* spp. are the dominant species above 1500 m. These mark the western limit of the Amazonian flora influence. *Ceroxylon* spp. are beyond this limit.

3.6 Secondary Forest

All disturbances of forest ecosystems (treefall gaps, storms, fluvial dynamics, deforestation for timber extraction, shifting cultivation, industrial plantations, or pastures) lead to a succession which is assumed to restore the floristic and architectural characteristics of the former ecosystem. The restoration of the forest becomes impossible when the deforested area is very extended and maintained for a long time.

Several arborescent species are frequently found in open areas and secondary forest in uplands. Pastures are invaded by *Orbignya phalerata* in the southeastern region (Brazil) and *Scheelea princeps* in Madre de Dios (Peru). *Astrocaryum chambira*, *Oenocarpus bacaba*, *Maximiliana maripa*, and *Socratea exorrhiza* form dense populations in deforested areas, while they occur with low density in the neighboring forests. The high density of these species in open areas and secondary forest is probably due to the fact that the adult palms are conserved when the forest is cut down because most of them are useful species; moreover, several of them are not seriously affected by burning. In old secondary forest, understory species become established when the canopy is already well developed. Only a few species, such as *Bactris simplifrons*, are occasionally found under the pioneer trees (*Cecropia* spp., *Bellucia* spp., *Byrsonima* spp., *Goupia glabra*, *Vismia* spp.).

On poor, white-sandy soils the vegetation is exceptionally cut down. Palms are in very low density on orthic podzol, while they form a large community on the irregularly waterlogged gleyic podzol. These latter soils support open-canopied forests where *Jessenia bataua* subsp. *bataua* is particularly abundant. After deforestation, *Mauritia flexuosa* and *Socratea exorrhiza* are the first to germinate in the open area. Seedlings and juveniles of *J. bataua* as well as those of many understory species are established only when a first foliage cover of pioneer trees is developed. Seedlings of *Mauritia flexuosa* may be found in relatively shady places (Richards 1952).

Open-canopied swamp forests are essentially composed of large palms such as *Euterpe precatoria*, *Mauritia flexuosa*, and *Socratea exorrhiza*. When the forest is cut down, seedlings and juveniles of these species restore the former vegetation. Only riparian species, such as *Astrocaryum jauari* and some *Bactris* spp., grow on the sediments newly deposited by whitewater rivers. The development of dense populations of *Astrocaryum carnosum*, *A.* cf. *chonta*, and *Phytelephas macrocarpa* takes place under a forest canopy dominated by *Ficus* spp. These medium-sized species are most abundant in disturbed habitats, as was observed in the upper Huallaga River valley in Peruvian Amazonia with *A. carnosum* and *P. macrocarpa*, both multistemmed palms. The formation of their dense populations is likely to be favored by the periodic disturbances of flooding and the superimposition of sylvigenetic processes as a result.

3.7 Savannahs

The coastal and subcoastal savannahs extend in the alluvial plain of the Guianan region. Several savannah types have been described according to soils and drainage conditions in Surinam (Heyligers 1963; Van Donselaar 1965, 1970a,b; Teunissen and Wildschut 1970), in Guyana (Cooper 1982), and in French Guiana (Hoock 1971). Cyperaceae and Poaceae form a discontinuous cover, which is usually burned at the dry season. The woody vegetation is composed of small shrubs of *Byrsonima verbascifolia*, and sparse treelets including *Byrsonima crassifolia* (Malpighiaceae), *Curatella americana* (Dilleniaceae), and *Clusia fockeana* (Clusiaceae) and *Humiria balsamifera* (Humiriaceae) on bleached soils. Palms are represented by *Acrocomia lasiospatha*, a single-stemmed species (Fig. 45), which is locally

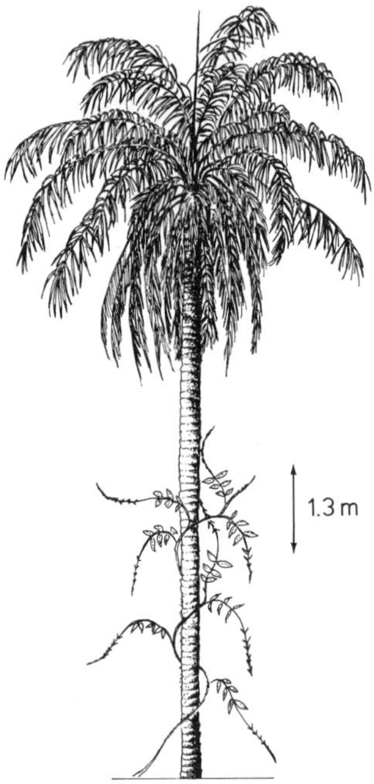

Fig. 45. *Acrocomia lasiospatha*, with the climbing *Desmoncus* sp.

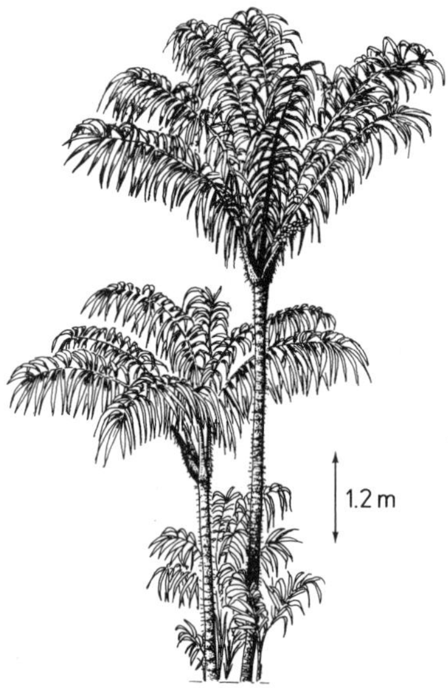

Fig. 46. *Astrocaryum vulgare*

abundant in the coastal belt, and *Astrocaryum vulgare*, a multistemmed species (Fig. 46), which is common on sandy soils in low forest patches and savannah borders. *Mauritia flexuosa* forms dense stands in inundated areas and gallery forests along creeks. Small *Bactris campestris* and *Desmoncus orthacanthos* form isolated thickets or dense populations on large areas.

Inland savannahs occur throughout the Amazon basin, generally on waterlogged, sandy soils. The northern savannahs are concentrated from the higher Rio Branco valley and in Roraima region to southern Surinam (Sipaliwini savannah). *Mauritia flexuosa* forms dense stands in this region. In the Tocantins River valley, the dominant palm is *Mauritiella aculeata*, a multistemmed species, which does not grow to more than 5 m in height in the savannah, while it reaches a height of up to 20 m in the contact with the neighboring forest.

Vast areas of savannah are found in the southern Amazonian regions of Brazil (Matto Grosso), Bolivia (Beni), and to a lesser extent in Peru (Madre de Dios). The most common palms are *Acrocomia totai*, *Scheelea princeps*, *Astrocaryum* spp. (Balslev and Moraes 1989), and *Syagrus sancona* which occurs in the islands of dry forest, and gallery forests.

3.8 Inselbergs

De Granville (1978) studied the vegetation on Tumuc Humac inselbergs, located at the junction of Brazil, Surinam, and French Guiana. The relative hygrometry strongly varies during the day on sunny days, falling to 40% at 14.00 h, with a peak around 98% at 2.00–3.00 h. On rainy days, the daily variation is negligible. The Venezuelan inselberg described by Blancaneaux and Pouyllau (1977) is characterized by a black granite surface, the temperature of which can reach 80 °C. The granite surface generally corresponds to cliffs. The soil is very thin or absent where the declivity is accentuated and accumulates in the fissures in the rock. The wind is an important factor conditioning the density of the vegetation.

On the slope of high declivity, the rock is covered with a discontinuous herbaceous vegetation classified as "rock-savannah". In the Tumuc Humac region, the "rock-savannah" is dominated by *Ischaemum guianense* (Poaceae) or *Pitcairnia geyskesii* (Bromeliaceae), or both species associated with *Chelonanthus uliginosus* (Gentianaceae), *Melampodium camphoratum*, *Riencourtia glomerata*, (Asteraceae), *Encyclia ionosma*, *Epidendrum nocturnum* (Orchidaceae), *Anthurium solitarium* (Araceae), *Trilepis kanukuensis* (Cyperaceae), *Ernestia blackii*, *E. rubra* (Melastomataceae), *Cassia saxatilis* (Caesalpiniaceae), *Portulaca sedifolia* (Portulacaceae), *Borreria latifolia*, *Sipanea pratensis* (Rubiaceae), *Stylosanthes hispida* (Papilionaceae); palms are absent. However, some species (*Bactris gastoniana*, *Bactris simplicifrons*) occur in low transition forests which are generally located at the top and at the base of cliffs and the steepest slopes (de Granville 1978). Species inventories carried out on many inselbergs in French Guiana have pointed out the high variability of the vegetation. Palms found in the low transition forests are also common in the neighboring terra firme forests; the multistemmed *Syagrus stratincola* is only known from the granite outcrops in the Guianas.

3.9 Conclusion

3.9.1 Palm Diversity

All Amazonian forest ecosystems are not equally rich in plants. The richest ecosystems are uncontestably the terra firme forests for trees, and this is true for palms as well. Palm species diversity peaks in terra firme forests of central and western Amazonian lowlands and decreases eastwards in the Brazilian state of Pará and Amapa Territory and in the Guianas, and westwards on the Andean mountain slopes. Palm diversity in open areas and secondary forest is very low. In western and central Amazonia, palm species richness is clearly higher in the terra firme forests than in the wetland forests. In eastern Amazonia, the differences of species richness between wetland and upland forest ecosystems are not so noticeable because of the relatively low diversity of palms in the latter (compare Tables 24 and 37).

3.9.2 Palm Density

Palms form large and diversified communities in most forest ecosystems. The high degree of dominance of one or a few species, which is the general case, is not limited to species-poor communities such as swamp forests. The palm community in some terra firme forests, which is characterized by both high density and species richness, is also largely dominated by a few species, as shown for instance by two inventories carried out in Peruvian Amazonia: one on well-drained, clayey soil where the palm richness reaches 29 species in 0.71 ha, with one species, *Lepidocaryum tessmannii*, representing more than 54.3% of the community, and the other on waterlogged, white sand where palm richness reaches 23 species on 0.27 ha with *L. tessmannii* and *J. bataua* representing together 79.9% of the community, respectively. In open areas, a few arborescent species reach very high density; this is the case of *Orbignya phalerata* in the southern part of the basin, which can form palm forests to the detriment of pastures.

3.9.3 Life Forms and Ecosystems

The understory of terra firme forests as well as of forests on periodically alluvial soils deposited by whitewater rivers is largely dominated by juveniles and adults of medium-sized and small species. The percentage of juveniles of arborescent species is higher in the understory of seasonal swamp forests on waterlogged soils than in terra firme and alluvial soil forests, and peaks in the permanently flooded swamp forests (Table 38).

Medium-sized palms are particularly abundant in the forests on alluvial soils. Small, single or multistemmed palms are frequent and abundant in all forest ecosystems. Of the 67 species known from French Guiana, 57 are medium-sized or small palms, 54 occupy forest understory (43 in terra firme forests), 40 of them are multistemmed. Species with creeping rhizomes, such as *Bactris cruegeriana*, *B. major*, *B. maraja*, are more frequent in open areas, savannahs, riverine vegetation, mangrove; this might be related to their ability to form dense clonal populations particularly adapted to fix sediments. Acaulescent species with small leaves are particularly frequent in swamp forests, in forests on periodically flooded alluvial soils, and on waterlogged sandy soils, mainly represented by *Geonoma acaulis* and

Table 38. Distribution of palm life forms in the 1–10-m stratum in Amazonian forest ecosystems

	Adults and juveniles of small and medium-sized species		Juveniles of arborescent species	
	No.	Percent	No.	Percent
Terra firme forests (ha):				
Ferrocaño (0.71)	5412	96.3	208	3.7
Copal (0.5)	1157	81.7	260	18.3
Forests on periodically flooded alluvial soils:				
Lower Ucayali (0.4)	1307	92.6	105	7.4
Upper Huallaga (0.4)	533	83.4	106	16.6
Seasonal swamp forests:				
Lower Ucayali (1)	1359	61.5	849	38.5
Upper Huallaga (1.6)	812	62.2	494	37.8
Swamp forests permanently flooded:				
Lower Ucayali (1)	479	49.0	498	51.0

G. macrostachys. This life form also occurs with lower frequency in terra firme on well-drained soils. This is also true of the prostrate-stemmed species. Subterranean-stemmed, large-leafed palms are found in terra firme forests on well-drained soils and on waterlogged sandy soils. Small, climbing palms of the genus *Desmoncus* are generally more developed in disturbed vegetation with open canopy (e.g., periodically flooded forests). Though they were not found in all forests surveyed because of their low density, they are also frequent in terra firme forests, generally located in clearings and in disturbed places.

Arborescent, multistemmed palms over 10 m in height are only encountered in open-canopied forests. This is the case of *Euterpe oleracea* in swamp forests, *Astrocaryum vulgare* and *Mauritiella aculeata* on sandy soils, and the riparian *Astrocaryum jauari*. The density of arborescent palms greater than 10 m in height is very low in the terra firme forests, high in the seasonal and permanent swamp forests, but rather low in the forests on periodically flooded alluvial soils. In spite of their low density in terra firme forests, diversity of large single-stemmed species is not significantly lower than in swampy vegetation they dominate. For instance, four of the six large, single-stemmed species commonly found in French Guiana, *Oenocarpus bacaba*, *Maximiliana maripa*, *Jessenia bataua* subsp. *oligocarpa*, and *Socratea exorrhiza*, form multispecific populations of low density in terra firme forests; the two latter also occur in seasonal swamps. The two other arborescent species, *Acrocomia lasiospatha* and *Mauritia flexuosa*, are dominant in coastal savannahs and swamps, respectively.

3.9.4 Key to Amazonian Forest Ecosystems Based on Their Palms

Amazonian forest ecosystems could be characterized by the floristic composition and structure of their palm communities, as attempted by the following key:

1. a Palms dominant in forest canopy (about 10 or more per 0.1 ha). [see 2]
1. b Palms not dominant in forest canopy (less than 4 per 0.1 ha). [see 9]
1. c Palms in savannahs or deforested area, not forming a palm forest. [see 11]
2. a An arborescent species strongly dominant. [see 3]
2. b Two to four arborescent species dominant. [see 6]
3. a A large, multistemmed species dominant. [see 4]
3. b A large, single-stemmed species dominant. [see 5]
4. a *Euterpe oleracea* strongly dominant: Seasonal or permanent swamp forests, or forests on periodically flooded alluvial soils under tidal influence; eastern Amazonia.
4. b *Astrocaryum jauari* strongly dominant: Riverside periodically flooded by white, black, and clear waters; throughout the basin, less frequent in the Guianas.
4. c *Mauritiella aculeata* strongly dominant: Irregularly waterlogged, sandy soils (gleyic podzol); throughout the basin.
4. d *Astrocaryum vulgare* dominant: Coastal dry forests on sandy soils; eastern Amazonia.
5. a *Mauritia flexuosa* strongly dominant: Permanently flooded swamp forests on peaty soils; throughout the basin.
5. b *Jessenia bataua* subsp. *bataua* strongly dominant: Irregularly waterlogged sandy soils (gleyic podzol); central and western Amazonia.
5. c *Orbignya phalerata* dominant: Secondary forest and pastures in unflooded areas; southern Amazonia.
5. d *Astrocaryum aculeatum* dominant: Deforested areas in central Amazonia.
5. e *Dictyocaryum lamarckianum* dominant: Montane forest above 1500 m in elevation; western limit of Amazonia.
6. a Dominant species are large, single-stemmed palms. [see 7]
6. b Dominant species are single and multistemmed palms. [see 8]
7. a *Jessenia bataua* subsp. *bataua*, *Euterpe precatoria*, and *Mauritia flexuosa* in the canopy: Seasonal swamp forests; central and western Amazonia.

7. b *Mauritia carana*, *Jessenia bataua* subsp. *bataua* in the canopy, *Euterpe catinga* in the intermediate strata: Waterlogged, sandy soils; central and western Amazonia.
7. c *Astrocaryum aculeatum* and *Maximiliana maripa* dominant: Secondary forest in unflooded areas; central Amazonia.
7. d *Astrocaryum chambira*, *Oenocarpus bacaba*, and *Maximiliana maripa* dominant: Secondary forest in unflooded areas; western Amazonia.
7. e *Scheelea princeps* and *Syagrus sancona*: Secondary forest and deforested areas; southern region.
7. f *Iriartea deltoidea*, *Jessenia bataua* subsp. *bataua*, *Mauritia flexuosa*: Seasonal swamp forest in eastern Andean piedmont.
8. a *Euterpe oleracea* and *Mauritia flexuosa*: Swamp under tidal influence; eastern Amazonia.
8. b *Mauritia flexuosa*, *Euterpe precatoria*, *Jessenia bataua* subsp. *bataua*, *Mauritiella aculeata*: Seasonal swamp forests; central Amazonia.
8. c *Maximiliana maripa* and *Euterpe oleracea* in the canopy, *Astrocaryum murumuru* in the understory: Forest on periodically flooded alluvial soils; eastern Amazonia.
9. a One to three medium-sized species strongly dominant between 1–10 m. [see 10]
9. b No dominant species in the understory: Terra firme forest; throughout the basin.
10. a *Astrocaryum* cf. *chonta*, *A. carnosum*, or *A. urostachys*, and/or *Phytelephas macrocarpa* abundant to 10 m in height, *Scheelea* spp. or *Maximiliana maripa* as arborescent palms: Forests on periodically flooded alluvial soils; central and western Amazonia.
10. b *Leopoldinia* spp. in the understory: Forest flooded by blackwater river; central Amazonia, along the Rio Negro valley.
10. c One or two small species dominant, with many others in lower density in the understory: Terra firme forest, throughout the basin.
11. a *Acrocomia lasiospatha*, single-stemmed species: Coastal savannah on sandy soils; eastern Amazonia.
11. b *Mauritiella aculeata*, multistemmed species: Savannah on sandy soils; eastern and northern Amazonia.
11. c *Barcella odora*, subacaulescent species: Savannah on sandy soils; northern Amazonia.

4 Palm Distribution as a Function of Soil Drainage, Topography, Forest Architecture and Dynamics, and Human Activities

4.1 Introduction

Chapter 3 discussed the distribution of palm species and the relative importance of the palm community in the main forest ecosystems of the Amazon basin. These ecosystems are built in function of water dynamics and flooding and fall into two main categories: (1) unflooded forests, including the terra firme forests on clayey, well-drained soils (acrisol, luvisol), and the forests on white-sandy soils which are either very dry (orthic podzol) or waterlogged after rainfall (gleyic podzol); (2) inundation forests, including the seasonal swamp forests on waterlogged soils (gleysol) irregularly flooded by small streams during the rainy season, the forests on alluvial soils (fluvisol) irregularly or periodically flooded by whitewater, the forests on clayey substratum periodically flooded by blackwater, and the swamp forests on organic soils (histosol) which are permanently flooded. Geological substratum and elevation largely determine inselberg flora, while rainfall and elevation are the dominant factors which condition montane flora.

It will be attempted here to distinguish between the effects of soil drainage, topography, and forest architecture and dynamics on the distribution of palm species and life forms. These three factors are closely related and together condition the palm distribution. The relative importance of each of them will be estimated from a comparison of several sites, defined by topography and soil, and forest architecture. Another important factor which cannot be excluded is man; human activities modify the architecture and dynamics of forest ecosystems and influence plant species distribution. Anthropic factors may have also played an important role in the formation and maintenance of most savannahs in the basin as well as at its periphery.

4.2 Soil Drainage and Topography

Water drainage depends on the characteristics of the soils (clayey or sandy), but is also conditioned by topography. Soils located in the bottom of upland valleys which collect water from slopes and plateaus are permanently water-

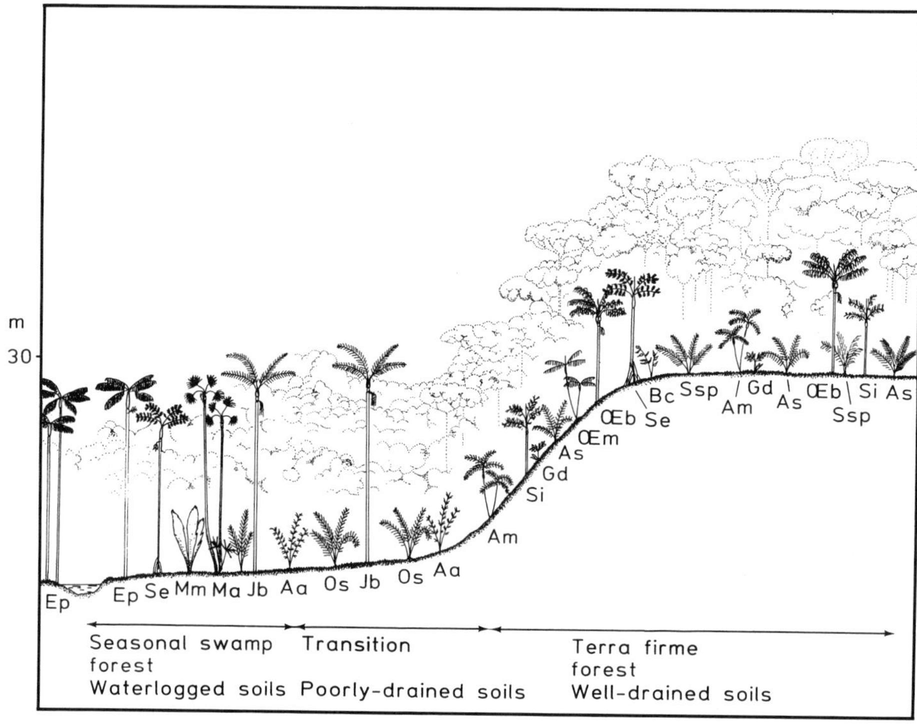

Fig. 47. Palms along a catena in a terra firme forest of central Amazonia (Kahn and Castro 1985). *Aa Astrocaryum acaule*; *Am A. gynacanthum*; *As A. sociale*; *Bc Bactris constanciae*; *Ep Euterpe precatoria*; *Gd Geonoma deversa*; *Jb Jessenia bataua* subsp. *bataua*; *Ma Mauritiella aculeata*; *Mm Manicaria martiana*; *Oeb Oenocarpus bacaba*; *Oem Oenocarpus minor*; *Os Orbignya spectabilis*; *Se Socratea exorrhiza*; *Si Syagrus inajai*; *Ssp Scheelea* spec. nov.

logged. Soils on slopes are generally well drained. On plateaus, vertical drainage of clayey soils is either good or limited, in which case they present characteristics of hydromorphy; only some sandy soils can be waterlogged in the upper part toward the surface, however. Topography also conditions flooding by the river; the highest terraces are never flooded, the intermediate are flooded for some time, and the lowest suffer inundation for several months.

Data from the former inventories are selected in order to illustrate the close relationship between soil drainage, flooding, and topography, and the subsequent effect on palm communities. Three cases are treated: (1) the variation in palm community along a catena from the plateau to the valley bottom, with three examples; (2) the variation of water dynamics in sandy soils in relation to topography; and (3) the distribution of one species in the different ecosystems from the lowest to the highest terraces in the lower Ucayali River valley.

Table 39. Distribution of palm species in relation to soils along two catenas in a forest of central Amazonia, near Manaus, Brazil. (Kahn and Castro 1985)

	WDS	PDS	WLS	Life form	Mature height (m)
Astrocaryum gynacanthum	+	–	–	Smt	3–10
Astrocaryum sociale	+	–	–	Subst	5
Bactris constanciae	+	–	–	Smt	3
Bactris elegans	+	–	–	Smt	3
Bactris humilis	+	–	–	Smt	3
Bactris (7 species)	+	–	–	Smt	2–4
Geonoma (5 species)	+	–	–	Smt	1–5
Oenocarpus bacaba	+	–	–	Ast	15
Scheelea sp.	+	–	–	Subst	5
Syagrus inajai	+	–	–	Mst	8–10
Geonoma sp.	+	+	–	Smt	1–5
Iriartella setigera	+	+	–	Smt	1–10
Oenocarpus minor	+	+	–	Mmt	8–10
Bactris simplicifrons	+	+	–	Smt	1–5
Euterpe precatoria	+	+	+	Ast	20–25
Socratea exorrhiza	+	+	+	Ast	15
Astrocaryum acaule	–	+	–	Subst	5
Jessenia bataua	–	+	+	Ast	20–25
Mauritiella aculeata	–	+	+	Amt	15
Orbignya spectabilis	–	+	+	Subst	5
Bactris sp.	–	–	+	Smt	1–5
Manicaria martiana	–	–	+	Mst	7

WDS: well-drained soils on plateau, crest, and slope; PDS: poorly drained soils in a transition zone from slope to valley bottom; WLS: waterlogged soils in valley bottom. Ast, Amt: arborescent single-stemmed, multistemmed; Mst, Mmt: medium-sized single-stemmed, multistemmed; Smt: small multistemmed; Subst: subterranean-stemmed and large-leafed.

4.2.1 Distribution of Palm Species Along a Catena

First Example. The distribution of palms was traced along two topographic sequences in a primary forest at the Experimental Station of Tropical Sylviculture (INPA), near Manaus, (Kahn and Castro 1985). The following conclusions were drawn (Fig. 47; Tables 39, 40):

- On well-drained soils, the vegetation includes arborescent palms which do not reach the forest canopy; arborescent multistemmed palms do not occur; the palm diversity in the understory (under 10 m in height) is particularly high, up to 17 species per 0.12-ha plot.
- On waterlogged soils which are flooded during the rainy season, arborescent palms are abundant in the forest canopy, reaching 30 m in height, and arborescent multistemmed palms are frequent; the palm diversity is low in the understory.

Table 40. Palm species richness per 0.12-ha plot in relation to topography in a forest of central Amazonia, near Manaus, Brazil. (Kahn and Castro 1985)

	Catena 1		Catena 2	
	Total species	Understory species	Total species	Understory species
Plateau	16	14	16	14
Crest	19	16	20	17
Slope	19	17	19	16
Transition zone	5	3	10	6
Valley bottom	7	3	6	2

- A transition zone between slope and valley bottom on poorly drained, sandy soils is particularly marked by the high density of two subterranean-stemmed palms with large leaves, *Astrocaryum acaule* and *Orbignya spectabilis*.

Second Example. A 26-m-long and 1-m-wide transect carried out in Piste de Saint Elie forest in French Guiana points out that the change in understory palm species is concomitant with change in soil drainage (Fig. 48). On well-drained soils, understory palms are represented by *Astrocaryum paramaca*, *Scheelea* spec. nov., both subterranean-stemmed, large-leafed species, and a seedling of *Jessenia bataua* subsp. *oligocarpa*; on poorly drained soils, there is only one palm species, the small, multistemmed *Geonoma oldemanii*, which forms a dense population among Marantaceae and Rapateaceae.

Third Example. The 0.18-ha plot inventoried in the Waki River valley in French Guiana (see Sect. 3.3.1) corresponded to a 180-m-long and 10-m-wide transect on plateau and slope. A contiguous 0.02-ha plot in the bottom valley on waterlogged soils included a different palm flora exclusively composed of *Euterpe oleracea* (Fig. 49). Of the 35 individuals counted on the area surveyed, ten were higher than 10 m, while only three individuals, two seedlings, and a single-stemmed juvenile of this species were counted on the 0.18 ha of well-drained soils.

4.2.2 Topography and Water Drainage in White-Sandy Soils

The drainage of water of white-sandy soils depends on the topographic location. The upper white-sand layer is highly permeable, but the underlying indurated hardpan is not, and retains water which cannot flow downwards. Thus, when a podzol is located on slopes or crests, the water can flow laterally, and the soil (orthic podzol) is dry even after heavy rainfall. When

Fig. 48. Understory palms along a 26-m transition from well-drained (*top*, *middle*) into waterlogged (*bottom*) soils at Piste de Saint Elie, French Guiana (de Granville 1978). *Ap Astrocaryum paramaca*; *Go Geonoma oldemanii*; *Jb Jessenia bataua* subsp. *oligocarpa*; *Ssp Scheelea* spec. nov.

the soil lies on a plateau or in a depression, water is retained by the hardpan and cannot flow laterally. The soil (gleyic podzol) is waterlogged after rainfall as a result. Herrera (1977) and Bongers et al. (1985) noted that the fluctuation of groundwater level in a podzol may rise toward the surface or

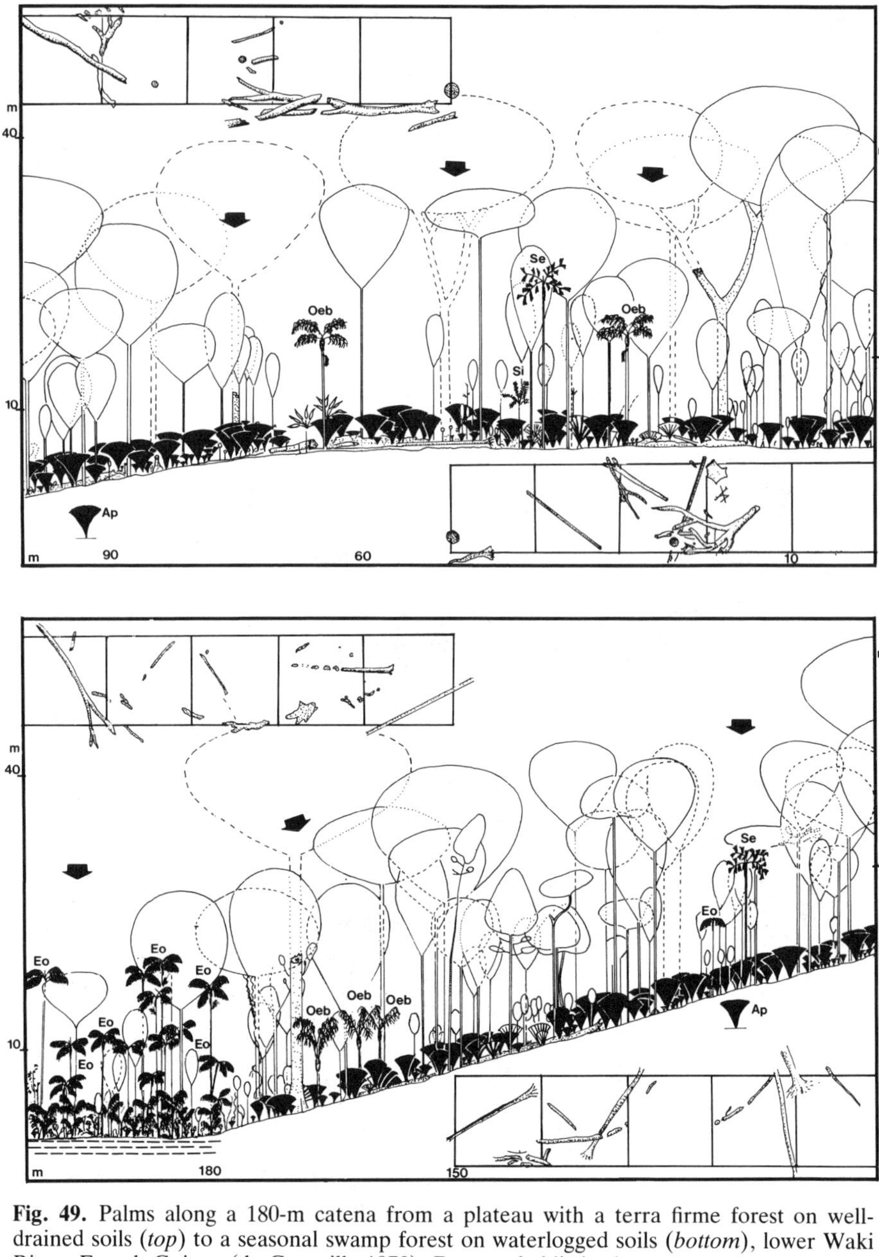

Fig. 49. Palms along a 180-m catena from a plateau with a terra firme forest on well-drained soils (*top*) to a seasonal swamp forest on waterlogged soils (*bottom*), lower Waki River, French Guiana (de Granville 1978). Recent chablis in the canopy are indicated by *arrow*. Adjacent canopy trees which are not rooted in the transect are marked by *dotted outline*. The *insets* (*top left* and *bottom right*) map the fallen trees. *Ap Astrocaryum paramaca*; *Eo Euterpe oleracea*; *Oeb Oenocarpus bacaba*; *Se Socratea exorrhiza*; *Si Syagus inajai*

Table 41. Mean density on 0.1 ha[a] of *Jessenia bataua* in the main forest ecosystems in the lower Ucayali River valley, Peru

Soils:	Gleyic podzol		Orthic[c] podzol	Humic gleysol		Orthic[f] acrisol	Eutric[e] fluvisol	Distric[d] histosol
Flooding:	Never		Never	Irregular by rain		Never	By river	Permanent
	A[b]	B[c]		C[c]	D[c]			
<1 m	209	510	11.5	119	5	21	0	0
1–10 m	94.5	141.5	20	28.9	15.5	27.5	0	0
>10 m	10.4	9.5	1	8	1.5	0	0	0

[a] Calculated from: [b]0.27 ha, [c]0.2 ha, [d]1 ha, [e]0.4 ha, [f]0.71 ha surveyed. A–B: two plots 11 km apart. C–D: in the same 1-ha plot of seasonal swamp forest; flooding is almost nil in C and well marked in D (small depressions).

above it. On the dry orthic podzol, the palm community is characteristized by low diversity and low density while on the waterlogged gleyic podzol the forest is characterized by high diversity and density (see Sect. 3.3.2, Tables 25 and 26). Several species generally found in seasonal swamp forests on a gleysol are also present on these waterlogged gleyic podzols.

4.2.3 The Distribution of *Jessenia bataua* subsp. *bataua*

In the lower Ucayali River valley (Peru), this species was inventoried in the six main forest ecosystems distributed from the river in permanently flooded swamps to the highest, unflooded terraces. The palm was found in four of the six forests surveyed (Table 41). Populations with highest densities occurred on irregularly waterlogged, white-sandy soils (gleyic podzol) located in unflooded places. The species was also abundant in seasonal swamp forests on permanently waterlogged soils (gleysol) located in upland valley bottoms irregularly flooded by small streams after rainfall. Population density of *J. bataua* subsp. *bataua* was very low on dry, white-sandy soils (orthic podzol). Only seedlings and juveniles were found on clayey, well-drained soils (orthic acrisol) in terra firme forests. No individuals were encountered in forests on periodically flooded alluvial soils (eutric fluvisol), or in swamp forests on permanently inundated, organic soils (histosol).

These results lead to the following conclusions: (1) Soils of both ecosystems where *J. bataua* subsp. *bataua* is abundant are characterized by being permanently or temporarily waterlogged and having a sandy texture. (2) *Jessenia bataua* subsp. *bataua* supports irregular, short flooding after rainfall (from some hours to a few days), but it is absent from places which are periodically flooded by a river. (3) The palm is only found as seedlings

Table 42. Architecture of forests (from plateau to valley bottom) and their palm communities. (Kahn 1986c)

	Lower Tocantins[a]	Central Amazonia[b]
Forest architecture		
DBH[c] (m)		
0.15–0.59	189	384
0.60–1.29	13	11
>1.3	5–6	–
Largest DBH (m)	2.5	1.25
Mean height of upper-canopy trees (m)	45	30
Largest size of chablis (ha)	0.12	0.06
Palm community		
Total area surveyed (ha)	10.56	1.2
Mean density of palms >1 m in height per ha	602	2122
Number of understory species	12	27
Number of subterranean-trunked species with large leaves	0	4
Density of subterranean-trunked species with large leaves	0	1345
Number of arborescent species in upland forest	5	2
Height reached by arborescent species	25–30	15

[a] 11 catenas, 0.96 ha each.
[b] 2 catenas, 0.6 ha each.
[c] Mean number of stems/ha calculated from 2.4 ha surveyed in the Tocantins forest, and from 1.44 ha surveyed and Prance et al. 's data (1976) on 1 ha in a central Amazonian forest.

and juveniles in terra firme forests; the low growth rate of this species (see Chap. 6) does not allow it to develop a trunk, even when growing in treefall clearings, as discussed below.

This species presents a large ecological range. *Jessenia bataua* subsp. *bataua* is presented as a terra firme forest palm in Ecuadorean Amazonia (Borgtoft Pederson and Balslev 1990); adults of *J. bataua* subsp. *oligocarpa* occur in seasonal swamp forest as well as in terra firme forest in French Guiana.

4.3 Forest Architecture and Dynamics

4.3.1 Sylvigenesis

Richards (1952) recognized three strata of trees (above approximately 15 m in height). Rollet (1974) lumped together architectural data from many forestry inventories over a large area and concluded that strata do not exist.

However, a vertical series of shrubs, medium-sized and large trees which "lack any potential for further expansion" (Hallé et al. 1978), is perceptible and can be analyzed in small areas. A forest is a mosaic of these small-sized eco-units, each with diverse architecture which represents a moment of sylvigenesis – aggrading, steady-state, or degrading phases (Oldeman 1983, 1990). Late aggrading, steady-state, and early degrading eco-units have a clearly layered architecture (Oldeman 1974; Kahn 1983). The surface of an eco-unit usually corresponds to natural treefalls in Amazonian terra firme forests (Oldeman 1978; Hartshorn 1978, 1980; Whitmore 1978, 1982; Denslow 1980; Brokaw 1982a; Riera 1986). The result of a treefall is a chablis. This ancient French word "denotes the uprooting of a tree, the fallen tree itself, the resulting opening in the forest and the debris on the forest floor" (Oldeman 1983, p. 142). The size of the chablis influences light availability, soil temperature, and humidity, and these factors determine colonizer composition (Vásquez-Yanes 1974; Holthuijzen and Boerboom 1982). Large chablis are most suitable for colonization by pioneer trees, as shown by Kramer's early experiments (1933) and other numerous studies summarized and discussed by Hartshorn (1978, 1980), Denslow (1980), Whitmore (1982, 1983), and Brokaw (1982b, 1985a,b). Sizes of larger chablis are related to the sizes of the largest trees.

4.3.2 A Comparison of Two Amazonian Terra Firme Forests

Amazonian terra firme forests can present contrasting architecture and palm composition (Table 42): (1) The forests of the lower Tocantins River valley are characterized by the great size of their trees in comparison with those in other Amazonian forests. Trees as high as 50 m are frequent, with DBH up to 2.5 m. Prominent species are *Bertholletia excelsa* (Lecythidaceae), *Astronium lecointei*, *Anacardium giganteum* (Anacardiaceae), and *Alexa grandiflora* (Papilionaceae). The "giant" species develop large crowns, and their death while standing creates the many complex chablis seen from the air. A complex chablis, up to 0.12 ha in area, is the result of successive falls of branches, each breaking small trees and shrubs in the understory, followed by the fall of the trunk, usually with a semi-regenerated crown. (2) The forests of central Amazonia are composed of smaller trees reaching 40 m in height and 1.3 m in DBH. The average height of the upper forest canopy is about 30 m. There are no particularly abundant species, although most frequent are *Manilkara surinamensis* (Sapotaceae), *Hevea guianensis* (Euphorbiaceae), *Scleronoma micranthum* (Bombacaceae), *Caryocar villosum* (Caryocaraceae), and *Dinizia excelsa* (Mimosaceae). While these species can develop large branches, "giant" trees overlooking these forests are rare. The falls of the biggest trees knock down smaller trees causing multiple chablis which do not exceed 0.06 ha in area.

There are notable differences in palm life forms between these forests. In the central Amazonian forest studied, all life forms are represented by the 27 understory species counted along the catena; four species with subterranean stem and large leaves are particularly abundant: *Astrocaryum sociale* and *Scheelea* spec. nov. on well-drained soils, *Astrocaryum acaule* and *Orbignya spectabilis* on poorly drained, sandy soils on the lowest part of the slope. Only a few individuals of *O. spectabilis* are encountered in the seasonal swamp forest at the valley bottom, in contrast to the very high density reached by this species in the other topographic position where the forest canopy is more closed. Such subterranean-stemmed palms with large leaves are totally absent from the Tocantins forests, where only 12 understory species – single and multistemmed palms – were found in 10.56 ha surveyed (Kahn 1986a).

Two arborescent palm species occur in terra firme forests of central Amazonia: *Oenocarpus bacaba* and *Socratea exorrhiza*. A third arborescent species, *Euterpe precatoria*, is only occasionally found in this community, although abundant in the neighboring seasonal swamp forest. Five arborescent species are frequent in the Tocantins terra firme forests: *Maximiliana maripa*, *Oenocarpus bacaba*, *O. distichus*, *Orbignya phalerata*, and *Socratea exorrhiza*. There they reach 25–30 m in height; in contrast, *O. bacaba* and *S. exorrhiza* do not exceed 15 m in height in central Amazonia. In both forests, these species were found below discontinuities of the upper canopy, indicating that they can grow only in chablis with enough light.

4.3.3 Arborescent Forms and Chablis

Light intensity levels required by forest palms are described by de Granville (1978). Two growth phases are distinguished (Fig. 50). The establishment growth (Tomlinson 1990), which corresponds to the elaboration of adult-sized leaves, requires progressively higher levels of light intensity. In the largest species, the juvenile plant keeps being acaulescent during this phase of the ontogenesis. Some tall species, however, such as *Euterpe precatoria* (see Table 56) and *Socratea exorrhiza* (Tomlinson 1990), produce a slender aerial stem. The second phase, which corresponds to the development of a trunk with adult diameter in both cases, requires a consistently high light intensity level. In fact, it occurs only when the palm stands in full sunlight, i.e., in chablis.

In the complex chablis of the Tocantins forests, of 0.10–0.12 ha in area, pioneer trees close gaps, whereas closing by development of the tree crowns from the border is negligible. Palm growth must keep pace with pioneer trees in order to compete effectively for light, and thereby receive the sunlight necessary for the elaboration of the stem (Fig. 51a). Arborescent palms are suitably adapted for regeneration in these sites.

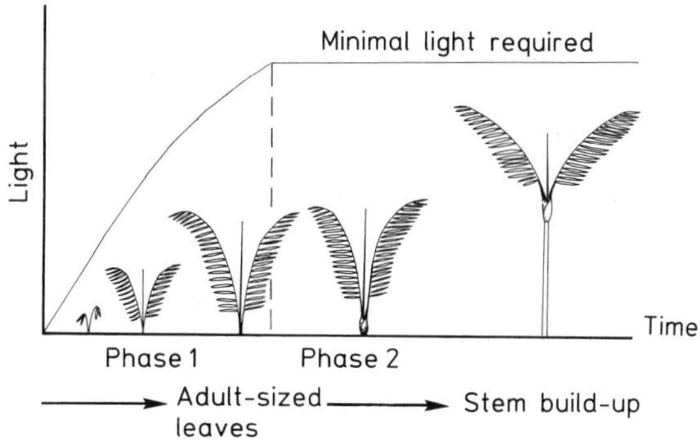

Fig. 50. Arborescent palm growth and light requirements. Minimal light required increases with leaf size (*phase 1*). With adult-sized leaves, the palm needs a consistently high light intensity level for building its stem (*phase 2*). (After de Granville 1978; Kahn 1986c)

In the smaller chablis of the central Amazonian forest (<0.06 ha), closing from the bordering tree crowns is relatively effective in producing a continuous layer of foliage. The light intensity that penetrates the understory layers consequently decreases. Arborescent palm species can reach the first growth phase, but they do not receive enough light for building their stem (Fig. 51b). Arborescent palms are found in the central Amazonian forest, but they occur on plateaus where chablis areas are greater because of a higher frequency of larger trees than on other topographic sites, and on crests where chablis are more frequent because of the wind and declivity effects.

In the transect of the lower Waki River valley, tall *Oenocarpus bacaba* and *Socratea exorrhiza* are found only in the clearings (Fig. 49).

The absence of arborescent, multistemmed palms from upland forest was noted and accounted for by de Granville (1978). All arborescent species require high light levels during the stage of stem growth and thus in the forest their regeneration tends to be restricted to chablis. As a chablis closes, the palm develops its first stem. It follows, then, that such chablis are not suitable for arborescent, multistemmed palms since the second axis initiated at the lower part of the stem will be shaded out, from the moment it emerges, by the pioneer trees then maturing in that chablis. This life form, however, can develop in the seasonal swamp forests, the canopy of which is more open with fewer trees; diffuse plus direct light is probably higher than in terra firme forests and sufficient to maintain stem growth.

a

b

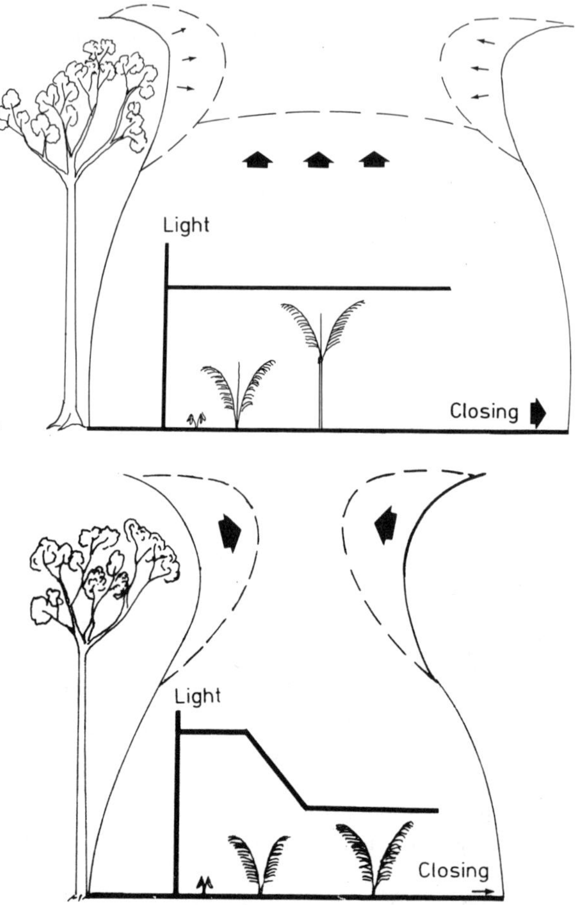

Fig. 51a,b. Arborescent palm growth in relation to gap size, gap closing, and available light. **a** Large gap (up to 0.12 ha) in the Tocantins forest. Pioneer trees close gaps, and closing by the tree crowns from the border is negligible. Palms receive the sunlight necessary for the elaboration of the stem. **b** Smaller gap (not exceeding 0.06 ha) in the central Amazonian forest. Closing by the tree crowns from the border is relatively effective, producing a continuous layer of foliages. Consequently, light intensity is decreasing in understory layers where palms do not receive enough light for building their stems. (Kahn 1986c)

4.4 Topography, Soil Drainage, and Forest Architecture and Dynamics

Topography influences soil drainage and also has a direct effect on forest architecture which determines the light intensity received by the understory plants. The variation of palm community size in relation to local topography and forest architecture and dynamics will be analyzed in the Tucuruí forest (see Sect. 3.3.1) and discussed with data from the central Amazonian forest in order to distinguish the respective influence of soil drainage and light on the palm community.

4.4.1 Topography and Forest Architecture and Dynamics

A practical method applied in the limited area of the eco-units (see Sect. 4.3.1) consists of counting the foliage layers of adult trees which lie above about 15 m in height (Figs. 52–54). Up to three foliage layers above that level have been observed in central and eastern Amazonian forests (Kahn 1987). An architecture with three upper foliage layers corresponds to a steady-state eco-unit. An architecture with one upper foliage layer corresponds to an aggrading eco-unit with development of pioneer trees. An architecture with two upper foliage layers may correspond to either an aggrading eco-unit at a later growth phase or a degrading eco-unit when upper trees are dying and their huge branches have fallen.

The frequency of chablis is generally higher on slopes with strong declivity, as discussed by Hartshorn (1978) and Oldeman (1974) in tropical America and by Guillaumet (1967) and Nierstraz (1975) in West Africa. The distribution of the largest trees is also influenced by topography, as shown by Huttel (1977) in the forests of Ivory Coast.

In the Tucuruí forest, 0.12-ha plots were disposed on plateaus, crests, slopes, and in depressions crossed by gullies with some patches of waterlogged soils. Declivity is nil on the plateau, smooth on the crest and in the lower part of the slope, and strong (>10%) on the slope. The frequency of steady-state eco-units with three layers of foliage above 15 m in height is higher on plateau and crest than on slope and depression (Table 43A). The largest trees are more frequent and basal area is significantly higher on plateau and crest ($m = 40.7\,m^2$; $s = 12.4$; $n = 16$) than on slope and depression ($m = 29.5\,m^2$; $s = 8.5$; $n = 16$). All the plots located on slopes present basal areas which are less than the mean ($35.1\,m^2/ha$), while 7 of the 16 plots located on plateau and crest present basal areas larger than the mean (Table 43B,C).

4.4.2 Palms and Forest Architecture and Dynamics

For this analysis a total of 32 plots (0.12 ha each) were surveyed: 14 plots displayed eco-units with three layers of foliage above 15 m, 12 had two layers, and six plots had just one layer above 15 m, representing a cumulative area of 1.68, 1.44, and 0.72 ha, respectively. Only the three species

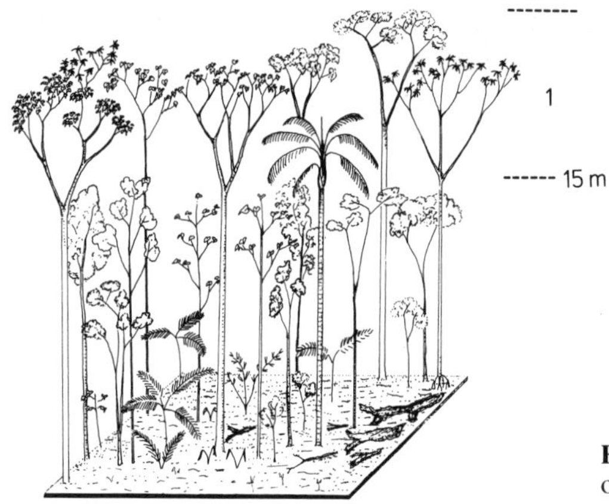

Fig. 52. Eco-unit with only one layer of tree foliage above 15 m in height

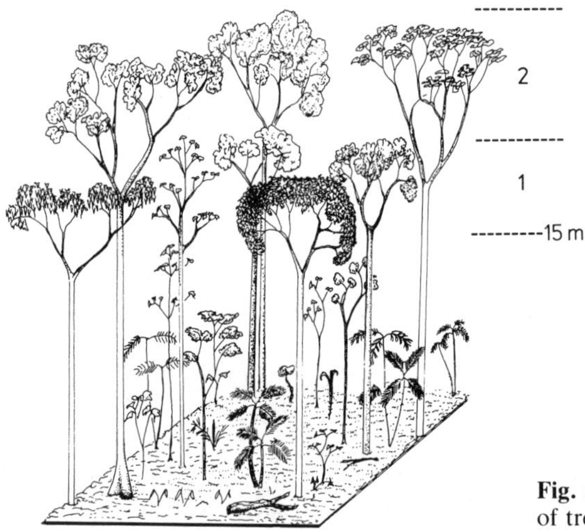

Fig. 53. Eco-unit with two layers of tree foliage above 15 m in height

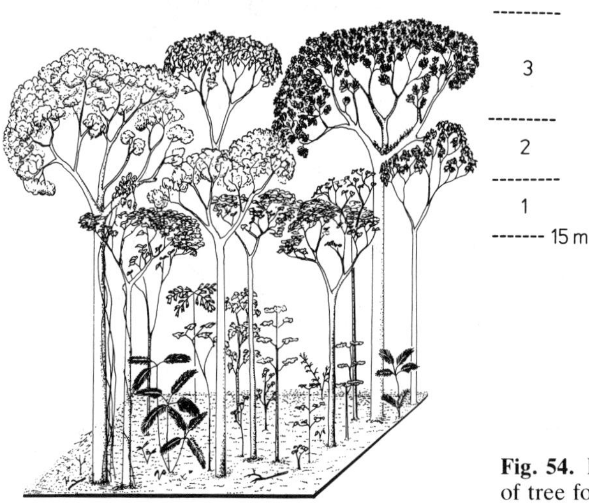

Fig. 54. Eco-unit with three layers of tree foliage above 15 m in height

Table 43. Forest architecture and topography in Tucuruí forest, Brazil. (Kahn 1987)

	Plateau	Crest	Slope	Depression	M	s
A) Frequency of plots with 1, 2, or 3 foliage layers above the 15-m level at each topographic site (n/8)						
1 layer	1	0	2	3		
2 layers	3	3	3	3		
3 layers	4	5	3	2		
B) Distribution of trees in diametric classes in relation to topography[a]						
0.15–0.39 m	187	194	204	180		
0.40–0.79 m	53	52	42	34		
0.80–1.19 m	9	6	4	2		
1.20–1.59 m	2	3	0	4		
>1.60 m	1	2	0	0		
C) Basal area[b]						
	32.26	33.11	23.68	37.05	31.53	5.63
	28.82	57.45	27.67	21.89	33.95	15.95
	42.07	32.59	34.14	54.21	40.75	9.89
	31.45	38.43	33.75	33.71	34.34	2.93
	33.29	58.85	21.09	21.74	33.74	17.65
	59.08	25.82	31.97	24.98	35.46	13.90
	32.93	31.74	31.08	27.12	30.72	2.52
	59.47	53.05	29.08	19.32	40.23	19.11
M	39.92	41.38	29.06	30.00	35.09	
s	12.54	13.03	4.70	11.53	11.88	

[a] On 0.96 ha – 8, 0.12-ha plots – at each topographic site.
[b] m^2/ha from DBH > 1.5 cm, calculated from the same plots. M: mean; s: standard deviation.

Table 44. Densities (per ha) of three palm species in relation to forest architecture (one, two, or three layers of tree foliage above 15 m), Tucuruí forest. (Kahn 1987)

	Forest architecture		
	One layer	Two layers	Three layers
No. of plots	6	12	14
Cumulative area (ha)	0.72	1.44	1.68
Astrocaryum gynacanthum	340	342	264
Bactris humilis	50	76	80
Oenocarpus bacaba	78	119	178

found in all 32 plots were considered (Table 44). The density of *Astrocaryum gynacanthum* is lower under a well-developed canopy, but is similar under canopies of one and two layers of tree foliage. The density of *Bactris humilis* and *Oenocarpus bacaba* increases with increasing number of foliage layers.

The variation of the populations of the other four species which do not occur in all plots cannot be interpreted in relation to forest architecture. Nevertheless, it is worthy of note that *Bactris elegans* occurs in all 14 plots with a forest architecture of three foliage layers above the 15 m level, but not in all plots with less-developed architecture (11/12 under two, and 3/6 plots under one foliage layer).

4.4.3 Palms and Topography

Of the 12 palm species found in the total area surveyed in the Tucuruí forest, seven with high densities are considered because only these permit quantitative comparison between different topographic sites. They are *Astrocaryum gynacanthum*, *Bactris humilis*, and *Oenocarpus bacaba*, which were found in all 32 plots, and *Bactris elegans*, *Geonoma deversa*, *Maximiliana maripa*, and *Syagrus inajai*, less frequent (Table 45).

The cumulative density of the seven palm species is lower on the slope and in the depressions (Table 46). *Geonoma deversa*, however, is less dense on the plateau and peaks in the depressions. The density of *Astrocaryum gynacanthum* is higher under less-developed forest architecture. As a consequence, one would expect larger populations of this species on slopes and in the depressions where aggrading eco-units are more frequent. However, this is not the case. On the contrary, there are fewer individuals in these two topographic sites than on crest and plateau. This suggests that declivity determines the magnitude of palm populations, that is, the effect of topography dominates over the presumed effect of forest architecture. In the depressions, attenuation of the declivity and a relative openness of the canopy have a cooperative effect on the regeneration of *A. gynacanthum*. The relatively more frequent chablis in the depressions are due to the falling of large trees from the slopes. The density of *Bactris humilis* and *Oenocarpus bacaba* is lower under less-developed forest architecture (one

Table 45. Frequency of palm species in relation to topography[a], Tucuruí forest. (Kahn 1987)

	Plateau	Crest	Slope	Depression
Astrocaryum gynacanthum	8	8	8	8
Bactris elegans	8	8	8	4
Bactris humilis	8	8	8	8
Geonoma deversa	6	5	5	8
Maximiliana maripa	7	7	6	3
Oenocarpus bacaba	8	8	8	8
Syagrus inajai	8	6	4	3

[a] Eight plots surveyed in each topographic site.

Table 46. Density of palms in relation to topography[a], Tucuruí forest, lower Tocantins River valley, Brazil (Kahn 1987)

	Plateau	Crest	Slope	Depression
Astrocaryum gynacanthum	381	327	189	285
Bactris elegans	159	273	76	17
Bactris humilis	89	98	43	48
Geonoma deversa	49	97	94	219
Maximiliana maripa	67	49	48	17
Oenocarpus bacaba	189(2)[b]	189(1)[b]	90	62
Syagrus inajai	41	26	7	6
Total = 3235	975	1059	547	654
100%	30.1%	32.7%	16.9%	20.2%

[a] 0.96 ha per each topographic site consisting of 8, 0.12-ha plots.
[b] Number of palms over 10 m in height in parentheses.

and two foliage layers) as well as on slopes and depressions where these architectures are more frequent. The effects of declivity and forest architecture cannot be separated because they act in the same direction. The two effects are also concordant with respect to *Bactris elegans*; the species appears with peak density on the crests and in all plots with three canopy layers, which are themselves especially frequent on the crests.

4.4.4 Conclusion: Topography, Water, and Light

Comparing this analysis of the eastern Tucuruí forest with data from central Amazonia, we may now try to isolate the basic factors which influence density and distribution of palms. In the central Amazonian forest, the canopy with three layers of foliage over 15 m was observed only on the plateau; crest and slope were covered in less-developed architectures (Kahn 1983). The palm community shows a clear peak of density on crests (Table 47) and, in contrast to the situation in Tucuruí forest, almost equal density on plateaus as on slopes. Hence declivity appears to have no effect. The puzzle can only be solved by considering topographic factors in more detail. In the central Amazonian forest, the transition from the plateau to the slope is abrupt because of the strong declivity of the slope (35%), whereas in the Tucuruí forest this transition is relatively smooth and gradual, which allows the canopy to be well developed. Hence in the former case, the trees on crests are more exposed to wind, and treefalls are more frequent, resulting in an open forest canopy. The intensity of light reaching the understory is higher; firstly, more light penetrates vertically in the more frequent gaps of the canopy, secondly, light passes laterally through the rupture between

Table 47. Palm density in relation to local topography[a] in a forest of central Amazonia, near Manaus, Brazil

	Plateau	Crest	Slope
Astrocaryum gynacanthum	37	60	49
Astrocaryum sociale	198	206	162
Bactris (11 species)	65	99	34
Euterpe precatoria	11	18	31
Geonoma (6 species)	69	86	102
Iriartella setigera	1	16	15
Oenocarpus bacaba	211(2)[b]	231(2)[b]	205
Oenocarpus minor	6	30	24
Scheelea sp.	86	91	75
Socratea exorrhiza	0	3	15
Syagrus inajai	16	66	8
Total	700	906	720

[a] 0.24 ha surveyed per topographic site.
[b] Number of palms over 10 m in height in parentheses.

plateau and slope. This is corroborated by observations from Guianan forests of Galbao mounts. The high density of *Bactris gastoniana* around 400 m in elevation and *Hyospathe elegans* from 400 to 500 m (see Sect. 3.3.1) corresponds to an area highly disturbed by treefalls on a crest where trees are particularly exposed to wind. A higher light penetration favors regeneration and growth in the majority of palms species as has also been shown in forests of Costa Rica (Chazdon 1986a; Richards and Williamson 1975), Mexico (Martínez Ramos 1980; Piñero et al. 1986), and Panama (De Steven 1986). In this latter study, the density of *Oenocarpus mapora* is clearly lower in a mature forest than in a disturbed one, but the mature forest population presents more large, many-stemmed clones; the author concluded (p. 103): "Overall, the high population densities and high seedling occurrence in the disturbed forest site suggest that the palm is favored in windthrown areas, perhaps because clones can survive canopy fall despite some ramet death and subsequently show strong growth response to increased light levels."

The central Amazonian study emphasizes the importance of light for the palm populations. However, the frequency of palms on the slopes – where forest canopy is composed of one or two layers of tree foliage – is equal to that on the plateaus with their fully developed canopy. The effects of light and declivity seem to neutralize each other on the slopes. In the Tucuruí forest the size of the palm community is clearly smaller on slopes and in depressions where the less-developed canopy is more frequent and, hence, where understory plants receive a relatively high light intensity; the pattern of palm distribution here requires a different interpretation. The effect of declivity on lateral drainage is dominant and leads to a lower density of the majority of the palm species on slopes and, to a certain extent, in the depressions. This effect is not compensated for by the higher light intensities in the understory of these topographic sites.

The comparison of these studies thus supports the conclusion that the characteristics of palm communities in the terra firme forests on well-drained soils are largely dependent on the following two factors: (1) the declivity, which determines lateral drainage of the soil, (2) the forest architecture, which influences the intensity of the light that is received by understory plants. However, the two factors are mutually dependent. Local topography not only determines declivity, but also influences the frequency of treefalls, i.e., forest architecture and dynamics. It plays an immediate role in the modification of drainage and thus of water supply, and affecting forest architecture interferes on light intensity in the understory.

4.5 Human Activities

Humans influence species distribution in three major ways: (1) by carrying species with them when they migrate, (2) by managing forest ecosystems and favoring the relative density of some species, (3) by destroying forest ecosystems and favoring the disappearance of species and the colonization of disturbed ecosystems by pioneer communities as a result.

4.5.1 Palms Transported by Humans

Several useful of ornamental species have been transported by humans throughout tropical South America. As a result, new species have been introduced in Amazonia, and the distribution area of some native palms has been increased. The populations of *Bactris gasipaes* Humboldt, Bonpland & Kunth in Amazonia are bound to be related to human deforestation for shifting cultivation (Balick 1984; Clement 1988). The great extension of *Mauritia flexuosa* throughout and beyond the Amazon basin is likely to be the result of transport by humans. The presence of *Acrocomia lasiospatha* and *Orbignya phalerata* in the outskirts of Manaus is also due to human activities. Neither palm is found in the forest of central Amazonia. A similar case is *Astrocaryum aculeatum*, which is so abundant in the town of Manaus and in secondary vegetation in the outskirts but totally absent from the surrounding primary forests. *Euterpe oleracea*, the distribution area of which is restricted to the eastern part of the basin is frequent in Manaus and other cities of central Amazonia. In Iquitos it was introduced 40 years ago in a private garden in the center of the town, and is now cultivated in other places. The royal palms have been introduced in the Amazon basin as

Fig. 55. Natural regeneration of *Roystonea oleracea*, an introduced species in French Guiana (lower Approuague River)

ornamental plants. *Roystonea regia* (Humboldt, Bonpland & Kunth) Cook, native of Cuba, is now growing in Puerto Maldonado (Madre de Dios, Peru). *Roystonea oleracea* (Jacquin) Cook, native of the Caribbean islands and Venezuela, is cultivated in French Guiana, where it is likely to form natural stands; de Granville (1989) found juveniles and adults of this species in a *Euterpe oleracea* swamp along the lower Approuague River (Fig. 55). The coconut tree, *Cocos nucifera* L., grows along all tropical coasts. It is so abundant in Devil's Island, a few kilometers from the Guianan coast, that it is eliminating the native vegetation. This palm is cultivated in gardens of many Amazonian towns, from the Atlantic coast to the eastern Andean piedmont. The African oil palm, which was introduced to Brazil via slave trade three centuries ago, is now constituting natural stands as if it were a native species. *Raphia taedigera* might have been introduced to eastern Amazonia in the same way (Otedoh 1977): according to Uhl and Dransfield (1987), however, this species is likely to have occurred in America before the complete separation of Africa from South America.

4.5.2 Palms as Indicators of Human Activities

As concluded by Balée (1988, p. 52), "palms are a major feature of the vegetation of many undisturbed archeological sites in Amazonia". According to this author, *Astrocaryum vulgare*, *Elaeis oleifera*, *Acrocomia lasiospatha*, *Maximiliana maripa*, and *Orbignya phalerata* are often associated with archeological sites on well-drained soils. He classified the dense and extensive stands in the State of Maranhão as "anthropogenic palm forests": "Palms often represent a major component of the vegetation of (such) previously utilized forests in Amazonia. The adaptation of modern Amazonian Indians to palm forests, especially those dominated by the babassu palm (*Orbignya phalerata* Mart.), may represent no mere adaptation to nature, but rather an adaptation to the residue of other cultures, some of which have been long extinct" (Balée 1988, p. 47). In the southern part of eastern Amazonia, the babassu palm, which covers around 200 000 km^2 in Amazonia (May et al. 1985a), invades pastures and secondary vegetation. The presence of *Astrocaryum vulgare* (Wessels Boer 1965) and *Maximiliana maripa* (Schulz 1960; Pesce 1985; Balée 1988) is said to indicate previous human settlement. *Astrocaryum chambira*, which occurs in the primary forest in Peruvian Amazonia at low density, is particularly frequent and abundant in secondary vegetation. All these species are useful palms.

The coastal savannah in French Guiana is dominated by *Astrocaryum vulgare*, *Acrocomia lasiospatha*, and *Mauritia flexuosa*. One hypothesis is that this savannah, as well as those of the southern part of the basin in Brazil and Bolivia, is likely to be anthropogenic.

4.5.3 Palms, Humans, and Forest Architecture and Dynamics

Humans destroy the forest by building highways throughout the Amazon basin. Such roads enable forest products to be exploited as well as facilitating the expansion of shifting cultivation by people moving from northeastern and southern Brazil in search of better living conditions. The first result is the destruction of the forest (Prance 1986). Some species which are highly adapted to the extreme ecological conditions of open vegetation follow man in his progression and colonize the new deforested lands. One of these is the babassu palm, *Orbignya phalerata*. Pastures are invaded in spite of frequent, repeated burning; juvenile babassu, the stem of which starts growing with a positive geotropism to a certain depth, can survive forest burning (Anderson 1983; Anderson and Anderson 1985).

The palm survey carried out in the lower Tocantins River valley in 1981 before the inundation by damming shows that the babassu extends farther northwards on the western side of the river than on the eastern side (Fig. 56). In fact, this region of the Tocantins valley is in the northern part of the

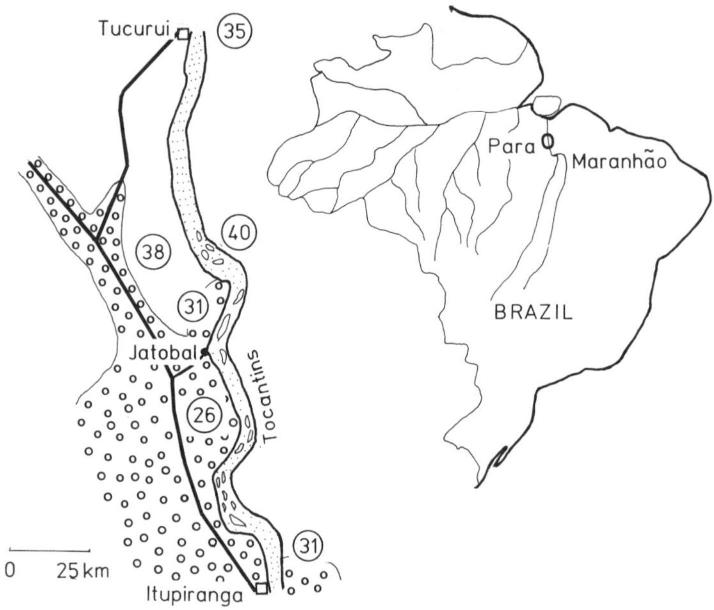

Fig. 56. Map of Brazil with enlargement of the lower Tocantins River valley and babassu distribution area in 1981 (*circles*). Basal area (calculated from DBH > 15 cm) is given for each survey area (26 m^2/ha inside babassu area, 31 m^2/ha at its northern limit, 35 to 40 m^2/ha in forests without babassu)

Fig. 57. *Orbignya phalerata*: abundance of seedlings in a terra firme forest in the lower Tocantins River valley

Fig. 58. *Orbignya phalerata*: abundance of acaulescent juveniles in the clearings of a terra firme forest in the lower Tocantins River valley. The numerous pinnate leaves shade the saplings of the largest trees, and as a result, inhibit their growth

species' distribution area. To the east of the river, it is not found north of the city of Itupiranga, whereas to the west of the river, it reaches the locality of Jatobal. The species has most likely expanded its distribution to the north along the transamazonian highway by colonizing deforested areas. It is also penetrating from the Tocantins banks into the forests along the small tributaries, first colonizing the seasonal swamp forests. From secondary forests, pastures, and seasonal swamp forests, the babassu penetrates into the neighboring upland primary forests by colonizing the large gaps which offer adequate ecological conditions for its full development, as has been described above for arborescent palms.

The data on density of this species strongly suggest that the palm is overrunning the forest: 170 palms (>1 m in height), all essentially located in chablis, and seven individuals (>15 m in height) were counted in 2.88 ha surveyed, and about 15 000 seedlings (<1 m in height) were found on 0.48 ha (Fig. 57). This species becomes abundant enough in the felled tree clearings to disrupt the development of upper-canopy tree saplings (Fig. 58). Denslow et al. (1991) presented evidence that in some forest palms the seedling density is low. The basal area of forests with no babassu is higher than in babassu overrun forests, (35 to 40 m^2/ha vs. 26 m^2/ha, respectively), and presents an intermediate value (31 m^2/ha) in forests located at the limit of its distribution area (Fig. 56).

These data lead us to the hypothesis that the forests of the Tocantins valley, which are characterized by large trees and thus by large chablis, are gradually transformed into a less-developed forest by two processes: (1) overrunning by the babassu palm and dense occupation of chablis, and (2) consequent disappearance of the tallest heliophilous trees which are unable to regenerate in chablis because their seedlings and saplings are shaded by large leaves of babassu, and crushed and smothered by palm litter. Hence, the floristic association – *Bertholletia excelsa* (Lecythidaceae) and *Orbignya phalerata* (Pires 1978) – may be unstable and correspond to the vanguard of babassu expansion with the correlated diminution of the density of the Brazil-nut tree as well as of others among the largest species.

In contrast, the arborescent palms, up to 25 m in height, which colonize the secondary forests and open vegetation in the outskirts of Manaus, are not found in the neighboring terra firme forests. Large trees are infrequent in these forests, and the resultant small chablis are closed by the crowns of the bordering trees, preventing the development of larger arborescent palms. Only palms up to 15 m in height are found here. This may explain the very low frequency or absence in these primary forests of the most common arborescent palms (*Astrocaryum aculeatum* and *Maximiliana maripa*) which occur in open areas in central Amazonia. *Oenocarpus bacaba* and *Socratea exorrhiza* are both moderate palms in the central Amazonian forests, while they are large palms in open vegetation as well as in the eastern Amazonian forests with large chablis.

5 Some Aspects of the Adaptive Radiation of Palms in Amazonia

5.1 Major Trends of the Adaptive Radiation of Palms

The amplitude of adaptive radiation of a taxon will be illustrated by its diversity of life forms and the range of its distribution in Amazonian forest ecosystems.

Cocoeae (12 genera), Areceae (7 genera), and Iriarteeae (6 genera) have representatives in all ecosystems. All three belong to the Arecoideae, which is the largest subfamily of Amazonian palms with 28 genera. Bactridinae (Cocoeae), with five genera, develops all the life forms found in the Amazonian palms and occupies all forest ecosystems. The genus *Astrocaryum* (Bactridinae) is the most diversified in life forms and provides a good example to illustrate palm adaptive radiation (Uhl and Dransfield 1987). This genus is clearly divided into two subgenera, *Pleiogynanthus* and *Monogynanthus* (Burret 1934). Most Amazonian species of the subgenus *Pleiogynanthus* are tall palms growing in open forests and deforested areas. Only one species, *A. acaule* (in fact, there are two other taxa, *A. huebneri* Burret and *A. giganteum* Barbosa Rodrigues, which are likely to fall into synonymy) is a subterranean-stemmed, large-leafed palm. There are no small Amazonian representatives in this subgenus and all Amazonian species present ragged leaves, i.e., leaves with pinnae oriented in several directions from the rachis. Most Amazonian species of the subgenus *Monogynanthus* are found in the understory of most forest ecosystems. They are medium-sized, large-leafed (5–7 m), or small, short-leafed (2–3 m) and slender-stemmed, or subterranean-stemmed, large-leafed palms. There are no tall palms in this subgenus except for *A. rodriguesii*, which can reach up to 20 m in height. All species develop leaves with the pinnae disposed in one plane. The other genera of Bactridinae are not so diversified in life forms in the Amazonian forests: *Bactris* and *Aiphanes* are small, single- or multistemmed species; *Acrocomia* are tall, single-stemmed species; *Desmoncus* develops only the climbing form. The genus *Oenocarpus* (Euterpeinae) has representatives in all ecosystems but develops only four life forms – arborescent or medium-sized, both single- or multistemmed. The other groups produce only one or a few life forms. Calamoideae are arborescent or medium-sized, single- or multistemmed palms (*Mauritia*,

Table 48. Life form diversity of palm genera in Amazonia

	Life form								
	Ast	Amt	Mst	Mmt	Sst	Smt	Subst	Ac	Cl
Acrocomia	+	−	−	−	−	−	−	−	−
Aiphanes	−	−	−	−	−	+	−	+	−
Ammandra	−	−	−	−	−	+[a]	−	−	−
Aphandra	−	−	+	−	−	−	−	−	−
Asterogyne	−	−	−	−	+	−	−	−	−
Astrocaryum	+	+	+	+	+	+	+	−	−
Attalea	+	−	−	−	−	−	−	−	−
Bactris	−	−	−	−	−	+	−	+	−
Barcella	−	−	−	−	−	−	−	+	−
Catoblastus	−	−	−	−	−	+[b]	−	−	−
Chamaedorea	−	−	−	−	+	+	−	−	−
Chelyocarpus	−	−	+	+	+[a]	−	−	−	−
Desmoncus	−	−	−	−	−	−	−	−	+
Dictyocaryum	+[b]	−	−	−	−	−	−	−	−
Elaeis	−	−	+[a]	−	−	−	−	−	−
Euterpe	+	+	+	−	−	−	−	−	−
Geonoma	−	−	−	−	+	+	−	+	−
Hyospathe	−	−	−	−	−	+	−	−	−
Iriartea	+	−	−	−	−	−	−	−	−
Iriartella	−	−	−	−	−	+[b]	−	−	−
Itaya	−	−	+	−	−	−	−	−	−
Jessenia	+	−	−	−	−	−	−	−	−
Leopoldinia	−	−	+	+	−	+	−	−	−
Lepidocaryum	−	−	−	−	−	+	−	−	−
Manicaria	−	−	+	+	−	−	−	−	−
Mauritia	+	−	−	−	−	−	−	−	−
Mauritiella	−	+	−	+	−	−	−	−	−
Maximiliana	+	−	−	−	−	−	−	−	−
Oenocarpus	+	+	+	+	−	−	−	−	−
Orbignya	+	−	−	−	−	−	+	−	−
Pholidostachys	−	−	−	−	+	−	−	−	−
Phytelephas	−	−	−	+	−	−	−	−	−
Prestoea	−	−	−	+	−	+	−	−	−
Raphia	−	−	−	+	−	−	−	−	−
Scheelea	+	−	−	−	−	−	+	−	−
Socratea	+	−	+[b]	−	−	−	−	−	−
Syagrus	+	−	+	+	−	−	−	−	−
Wendlandiella	−	−	−	−	−	+	−	−	−
Wettinia	−	−	+	+	−	−	−	−	−

Ast: Large, single-stemmed; Amt: Large, multistemmed; Mst: medium-sized, single-stemmed; Mmt: medium-sized, multistemmed; Sst: small, single-stemmed; Smt: small, multistemmed; Subst: subterranean-stemmed with large leaves; Ac: subacaulescent with small leaves; Cl: climbing.

[a] With prostrate stem.

[b] Production of stolons.

Mauritiella, *Raphia*) in swampy areas and small, multistemmed palms (*Lepidocaryum*) on usually well-drained soils in terra firme forests. Coryphoideae (*Chelyocarpus* and *Itaya*) and Phytelephantoideae (*Phytelephas*, *Ammandra*, *Aphandra*) are medium-sized or small, single- or multistemmed palms.

The most frequent life forms are the arborescent, single-stemmed palms developed by 14 genera; the small, multistemmed palms are found in 13 genera; and the medium, single- and medium, multistemmed palms in 12 and 11 genera, respectively (Table 48). In fact, there are genera specialized in large palms (*Acrocomia*, *Attalea*, *Dictyocaryum*, *Iriartea*, *Jessenia*, *Mauritia*, *Maximiliana*, *Oenocarpus*), in medium-sized palms (*Aphandra*, *Itaya*, *Manicaria*, *Phytelephas*, *Wettinia*), or in small palms (*Aiphanes*, *Ammandra*, *Asterogyne*, *Bactris*, *Barcella*, *Catoblastus*, *Chamaedorea*, *Geonoma*, *Hyospathe*, *Iriartella*, *Pholidostachys*, *Wendlandiella*). It is also true that many genera are represented by only one or a few species in Amazonia.

5.2 Some Phylogenic Aspects in Life Forms

To gain a better understanding of the adaptive radiation trends of palms in the Amazonian forest ecosystems, relationships between small and arborescent forms will be discussed: miniaturization or gigantism (Hallé and Oldeman 1970), and modification of one organ.

5.2.1 Miniaturization or Gigantism

Genera *Bactris* and *Geonoma* are mainly composed of dwarf species. Both are most diversified in the understory of terra firme forests where light intensity is weak. The smallest Amazonian species of *Astrocaryum* and *Oenocarpus* are also found in the understory of terra firme forests. Chazdon (1986b, p. 98) concluded that "small plants with low biomass investments in stem tissue and low leaf-support costs at the crown level have reduced metabolic costs and can grow in relatively more shaded conditions".

A morphological relationship of miniaturization or gigantism can be established between arborescent and small species. Small palms are considered to be miniatures of the larger species. The process of miniaturization or gigantism, which modifies all the parts of the plant, involves a complex evolutionary process, usually linking genera at the subtribe or subfamily levels, or species within some genera (Table 49A). This process is concomitant with a simplification of the blade morphology. The leaves of many

Table 49. Relationships between large and small life forms

A) Miniaturization or gigantism: large vs. medium-sized and small palms

Subfamily level: *Mauritia*, *Mauritiella* vs. *Lepidocaryum*.
Subtribe level: Bactridinae – *Astrocaryum*, *Acrocomia* vs. *Bactris*, *Aiphanes*. Elaeidinae – *Elaeis* vs. *Barcella*. Euterpeinae – *Euterpe*, *Jessenia*, and *Oenocarpus* vs. *Hyospathe*. Iriarteinae – *Dictyocaryum*, *Iriartea*, and *Socratea* vs. *Iriartella*
Genus level: *Astrocaryum* – *A. Jauari* vs. *A. gynacanthum*. *Oenocarpus* – *O. bacaba*, *O. mapora* vs. *O. balickii*, *O. minor*. *Prestoea* – *P. schultzeana* vs. *P. asplundii*. *Socratea* – *S. exorrhiza* vs. *S. salazarii*. *Syagrus* – *S. sancona* vs. *S. inajai*

B) Aerial vs. subterranean-stemmed species

Astrocaryum: (*Pleiogynanthus*) *A. aculeatum* vs. *A. acaule*; (*Monogynanthus*) *A. javarense*, *A. macrocalyx*, *A. sciophilum* vs. *A. paramaca*, *A. sociale*. *Orbignya*: *O. phalerata* vs. *O. polysticha*, *O. sagotii*, *O. spectabilis*. Scheelea: *S. brachyclada* vs. *S. insignis*

small palms present an entire blade, a juvenile character which classifies them as neotenic organisms (see Sect. 2.3.). For instance, *Syagrus inajai* develops entire leaves, up to 2.5 m in length, which become finely pinnate only when the stem is being elaborated, and then persist pinnate (Fig. 59), as is commonly observed in Brazilian populations of Pará and Amazônas. At the northern limit of the distribution area of the species, in the region of Saül, French Guiana, a population presents a different behavior: the stem is elaborated, up to 3–4 m in height, when leaves are still entire. The leaves later become pinnate and the palm starts flowering and fruiting. Finally, either the palm keeps producing pinnate leaves or it starts producing entire or sometimes partially pinnate leaves again (de Granville 1977). The succession of adult and juvenile phases, which is characteristic of this population, suggests a transformation into a neotenic organism. Adults of *Syagrus smithii* develop entire leaves in Colombian Amazonia (Galeano 1991) and pinnate leaves in other regions of the basin.

5.2.2 Modification of One Organ

On the other hand, arborescent, large-stemmed species, as well as the subterranean-stemmed species, develop large leaves. The transition from one form into the other involves only one organ, i.e., the reduction or development of the trunk. Positive geotropism is not only a characteristic of subterranean-stemmed palms. It corresponds to the establishment growth defined as the "saxophone" type (Tomlinson 1990). Several palms, such as the tall *Orbignya phalerata* (Anderson 1983), the medium-sized *Astrocaryum sciophilum* (Sist 1989b), and the small *Bactris gastoniana* (Fig. 60) present a first geotropic phase followed by an apogeotropic growth (Dransfield

Fig. 59a,b. *Syagrus inajai.* **a** Juvenile entire leaves; the youngest leaf starts dividing into pinnae. **b** Adult, medium-sized palm with ragged, finely pinnate leaves

1978) and the building of an aerial stem as a result. In Amazonia, the subterranean-stemmed habit is found in only three genera, *Astrocaryum*, *Orbignya*, and *Scheelea*, which all also develop the aerial-trunked habit (Table 49B).

The subterranean-stemmed forms dominate the understory of the central Amazonian forest studied in the former chapters, which is characterized by small trees and small chablis. When gaps in the canopy close from the tree crowns of the border, palms do not receive the required light intensity to build a stem. They remain at a low light intensity level and can only elaborate a subterranean, short stem. Such life forms do not occur in the eastern Tocantins forest, which is characterized by "giant" trees and resultant large and complex chablis. In forests with intermediate-sized trees some are present but not abundant. These include *Orbignya polysticha* in Peruvian and French Guianan forests, and *Astrocaryum paramaca*, *Orbignya sagotii*, and *Scheelea* sp. in French Guiana forests (de Granville 1976, 1978, 1989; Sist 1989a).

The subterranean-stemmed form with large leaves must be understood as the result of an adaptive process, probably related to precocious flowering. Some individuals of arborescent species are quite often observed flowering

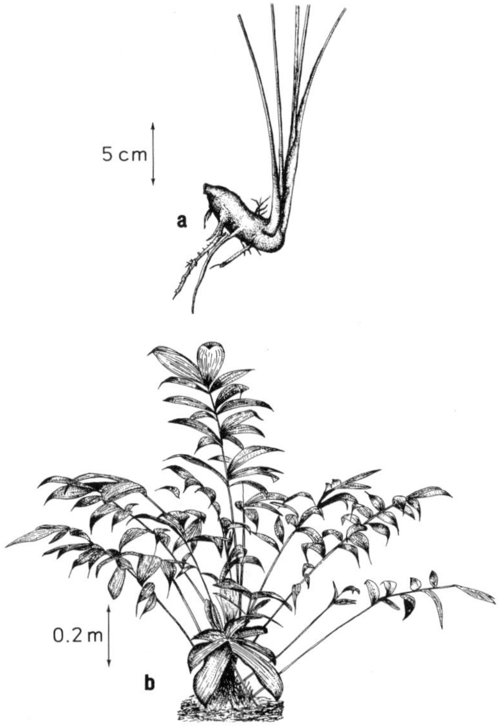

Fig. 60a,b. *Bactris gastoniana*. **a** Geotropic growth of the stem. **b** This small palm, still acaulescent, can produce a slender stem, up to 3 m in height

when they are still acaulescent. These plants will produce more fruits during their lifetime than those which start flowering later when a trunk is elaborated. The differential reproductive rate will progressively increase the frequency of the precocious flowering character in the population and the trend will be to reduce adult height. In this respect, the subterranean stem is not a direct (adaptive) response to decreasing light intensity towards the forest floor, but rather secondarily correlated (and adapted) to this gradient. Chazdon's (1986b, p. 98) conclusion about small-leafed palm adaptation is also true for subterranean-stemmed, large-leafed species: "the ability of a plant to reproduce while still at a small size creates the evolutionary potential for shade adaptation".

In tropical rain forests, the abundance of subterranean stems must be related to light intensity, which is influenced by forest architecture and dynamics. However, this life form also occurs in very open and drier vegetation such as the campos cerrados in southern Brazil (Bondar 1964; Medeiros-Costa and Panizza 1983). In this case, subterranean-stemmed palms receive constant intensive light. The main ecological constraint is water availability, and the subterranean stem is probably an adaptation to dry conditions, as suggested by Rawitscher and Rachid (1946). Moreover, "the habit has a survival value when in dry seasons fire may start and burn off the superficial vegetation leaving the underground stems unharmed; with their capital reserves they can soon replace the scorched leaves" (Corner 1966, p. 91). Thus, the adaptive value of the subterranean stem in palms is related to habitats.

5.3 Adaptive Value of Some Leaf and Root Structures

Palm structures are described in Chapter 2. The adaptive value of distichous crown, pneumatophores, and stilt-root structures has been questioned by several authors. We present here a review of this discussion.

5.3.1 Distichous Crown

"The significance of the distichous habit is not understood" concluded Dransfield (1978). The arborescent, single-stemmed palms *Oenocarpus bacaba*, with several leaf ranks, and *O. distichus*, with a distichous crown, occur in the forest of the lower Tocantins River valley, Brazil (Kahn 1986a). Adults of both species are found in the same stratum (15–25 m) under discontinuities of the forest canopy. The ecological conditions (soil and light) are also identical for the "abnormal" distichous *Mauritia flexuosa* found near Iquitos and for its "normal" congeners (Kahn 1988a). These facts suggest that the distichous crown arrangement in palms does not have adaptive value, but much of the biology of these palms remains unknown.

5.3.2 Pneumatophores

The ecological significance of these structures is obvious. Palms with pneumatophores are able to grow in anaerobic conditions in swampy areas. *Mauritia flexuosa*, *Euterpe oleracea*, and *E. precatoria*, which are the most common palms on permanently inundated organic soils, develop pneumatophores. These species are also found in open areas on well-drained soils. The opposite is not true: palms of terra firme forests on well-drained soils never occur on waterlogged soils or in swampy areas, except for some stilt-rooted species. Most Amazonian palms do not develop pneumatophores; they have small, spiny, white pneumatorhizae characterized by aerenchyma (see Sect. 2.2.5). Many of them are terra firme forest palms.

Little is known about the physiology of the pneumatorhizae. They may allow several species, such as *Astrocaryum* cf. *chonta*, *Elaeis oleifera*, *Jessenia bataua*, to support waterlogging and short flooding.

5.3.3 Stilt Roots

The stilt-rooted palms found in Amazonia are mainly represented by the Iriarteeae (Arecoideae). Henderson (1990) presents a complete review of the literature on the topic and concludes (p. 19): "Explanations have varied from adaptation to inundated soils, adaptation to exploit light gaps, physiological adaptation, adaptation to germination on steep and unstable slopes, and adaptation to flattening by falling branches" (Bouillenne 1924; Corner 1966; Jeník 1973; Givnish and Hartshorn's discussion in Dransfield 1978; Yeaton 1979; Bodley and Benson 1980; Schatz et al. 1985).

Wallace (1853) described *Socratea exorrhiza* as a swampy soil species, and Bouillenne (1924) found it in forests on periodically flooded alluvial soils (Brazilian várzea). In fact, it occurs in all forest ecosystems (Dransfield 1978; Kahn and Castro 1985). In Amazonia, it is clearly a pioneer species which colonizes the clearings in terra firme forests. Its higher frequency in the swamp forests is related to the open canopy of these. The stilt roots grow very rapidly (Vilhena et al. 1984) and upon reaching the soil, they branch into a multitude of fine roots. Each stilt root can produce a dense absorbing system because of its large diameter (up to 7 cm in adults). The early and rapid development of the stilt root permits the species to colonize the soil densily and bestows upon it its pioneer status (Kahn 1977, 1982). Moreover, the roots are covered with many small, white pneumatorhizae. The system composed by the stilt root with pneumatorhizae above the ground and the dense set of absorbing roots under the ground functions as a pneumatophore and, as a result, the species is able to grow on poorly drained soils. The pneumatorhizae on old stilt roots are hard, "woody" spines. They are bound to lose their breathing function, but can provide an effective protection against predators. This palm presents a very large range of distribution and occurs in the forests of Panama and Costa Rica. There it colonizes treefalls (Yeaton 1979), and its seedlings are unable to keep growing in low light conditions (Vandemeer et al. 1974). However, Hogan (1986) shows that this species does not require light gaps to grow in forests in Panama. Though contrasting with the other observations, this latter conclusion is not really surprising. Several species present different ecological behavior when they are located at the margin of their distribution area.

Iriartea deltoidea also develops large stilt roots; but these are produced later, when the plant has already made a stem several meters in height. The production of the first stilt roots takes place as the pinnae start dividing longitudinally into segments. This species is not so well identified a pioneer as is *S. exorrhiza*.

Stilt roots are also produced by the smallest species of Iriarteeae. In the Amazonian lowlands, *Socratea salazarii*, *Iriartella setigera*, *I. stenocarpa*, and *Wettinia augusta* grow in the understory of terra firme forests on well-drained soils, and *Catoblastus drudei* in seasonal swamp forests on water-logged soils. In the Andean piedmont *W. augusta* is usually found in seasonal swamp forests.

The fact that stilt roots are produced by all arborescent, medium-sized and small species of Iriarteeae and are found on well-drained as well as on waterlogged and flooded soils leads to understand the structure as an evolutive trend in this subfamily; the ability to produce stilt roots would therefore not be an adaptive character. Arborescent, stilt-rooted palms, however, can be related to gap colonization (i.e., pioneer status) as well as to swampy conditions; as functional organs, stilt roots allow these species to have a larger ecological range.

5.4 Overview of the Adaptive Radiation of Palms: Ecosystems, Life Forms, and Most Representative Species

Nature is not so abrupt as the words are; all is tendency. Most species are characteristic of one ecosystem in which they are particularly abundant. They can occasionally be found in other ecological conditions, in low density, however. Others are found in several ecosystems, and in each of them, they form representative populations; others present different ecological behavior from one region to another. A species can be classified as different life forms, tall in forest and medium-sized in inland savannah, for instance; and this turns into a Chinese puzzle when the wide ecological variability of one species superimposes with misunderstood taxonomy.

5.4.1 Terra Firme Forests

5.4.1.1 Arborescent Palms

Large: leaf length > 4 m; DBH > 20 cm – single-stemmed: *Astrocaryum aculeatum*, *Astrocaryum chambira*, *Iriartea deltoidea*, *Jessenia bataua* subsp. *bataua* (in Ecuador), *Jessenia bataua* subsp. *oligocarpa*, *Maximiliana maripa*, *Oenocarpus bacaba*, *O. distichus*, *Orbignya phalerata* – **multistemmed:** none.

Slender: leaf length < 4 m; DBH < 20 cm – single-stemmed: *Dictyocaryum ptariense*, *Socratea exorrhiza* – **multistemmed:** none.

5.4.1.2 Medium-Sized Palms

Large: leaf length > 4 m; DBH > 12 cm – single-stemmed: *Astrocaryum javarense*, *A. rodriguesii*, *A. sciophilum* – **multistemmed:** none.
Slender: leaf length < 4 m; DBH < 12 cm – single-stemmed: *Oenocarpus balickii*, *Socratea salazarii*, *Syagrus inajai*, *S. smithii*, *Wettinia maynensis* – **multistemmed:** *Oenocarpus minor*, *Wettinia augusta* (western lowlands).

5.4.1.3 Small Palms

Erect, single-stemmed: *Aiphanes deltoidea*, *Bactris oligocarpa*, *Chamaedorea pauciflora*, *Geonoma poeppigiana*, *G. triglochin*, *Pholidostachys synanthera* – **prostrate, single-stemmed:** *Chelyocarpus repens* – **erect, multistemmed:** *Astrocaryum gynacanthum*, *Bactris acanthocarpoides*, *B. acanthospatha*, *B. aubletiana*, *B. constanciae*, *B. humilis*, *B. elegans*, *B. pectinata*, *B. mitis*, *B. monticola* (east), *B. rhaphidacantha*, *B. simplicifrons*, *B. sphaerocarpa*, *Geonoma deversa*, *G. leptospadix*, *G. maxima*, *G. piscicauda*, *G. pycnostachys*, *G. stricta*, *G. spixiana*, *Hyospathe elegans*, *Iriartella setigera*, *I. stenocarpa* – **climbing, multistemmed:** *Desmoncus* – **acaulescent:** *Aiphanes ulei*, *Bactris gastoniana*, *Geonoma acaulis*, *G. poiteauana*.

5.4.1.4 Subterranean-Stemmed, Large-Leafed Palms

Astrocaryum paramaca, *Astrocaryum sociale*, *Orbignya polysticha*, *O. sagotii*, *Scheelea* spec. nov.

5.4.2 Forests on Periodically Flooded Alluvial Soils

5.4.2.1 Arborescent Palms

Large – single-stemmed: *Attalea tessmannii*, *Iriartea deltoidea*, *Orbignya phalerata*, *Scheelea bassleriana*, *S. brachyclada*, *S. cephalotes*, *S. stenorhyncha* – **multistemmed:** *Astrocaryum jauari* (river bank).
Slender – single-stemmed: *Euterpe precatoria*, *Socratea exorrhiza* – **multistemmed:** *Oenocarpus mapora*.

5.4.2.2 Medium-Sized Palms

Large – single-stemmed: *Astrocaryum* cf. *chonta* – **multistemmed:** *Astrocaryum carnosum*, *A. murumuru*, *A. urostachys*

Slender – erect, single-stemmed: *Chelyocarpus ulei*, *Itaya amicorum* – **erect to creeping, multistemmed:** *Phytelephas macrocarpa*.

5.4.2.3 Small Palms

Erect, single-stemmed: *Geonoma jussieuana* – **erect, multistemmed:** *Bactris bifida*, *B. cruegeriana*, *B. major* (restricted to areas under tidal influence), *Geonoma pycnostachys* – **climbing, multistemmed:** *Desmoncus* – **acaulescent:** *Geonoma acaulis*.

5.4.3 Forests Periodically Flooded by Blackwater

5.4.3.1 Arborescent Palms

Slender – multistemmed: *Astrocaryum jauari*.

5.4.3.2 Medium-Sized Palms

Slender – multistemmed: *Leopoldinia major*.

5.4.3.3 Small Palms

Erect, single-stemmed: *Leopoldinia pulchra* – **erect, multistemmed:** *Bactris concinna*, *Bactris maraja* – **climbing, multistemmed:** *Desmoncus* spp.

5.4.4 Swamp Forest on Organic, Permanently Flooded Soils

5.4.4.1 Arborescent Palms

Large – single-stemmed: *Mauritia flexuosa*.
Slender – single-stemmed: *Euterpe precatoria*, *Socratea exorrhiza* – **multistemmed:** *Euterpe oleracea*.

5.4.4.2 Medium-Sized Palms

Slender – multistemmed: *Oenocarpus mapora*.

5.4.4.3 Small Palms

Erect, multistemmed: *Bactris concinna*; **climbing, multistemmed:** *Desmoncus* spp.

5.4.5 Seasonal Swamp Forests on Waterlogged, Irregularly Flooded Soils

5.4.5.1 Arborescent Palms

Large – single-stemmed: *Jessenia bataua* subsp. *bataua*, *J. bataua* subsp. *oligocarpa*, *Mauritia flexuosa*.
Slender – single-stemmed: *Euterpe precatoria*, *Socratea exorrhiza*; **multistemmed:** *Euterpe oleracea*, *Mauritiella aculeata*.

5.4.5.2 Medium-Sized Palms

Large – single-stemmed: *Astrocaryum* cf. *chonta*, *Manicaria saccifera* – **creeping, single-stemmed:** *Elaeis oleifera* – **multistemmed:** *Astrocaryum murumuru*.
Slender – multistemmed: *Oenocarpus mapora*, *Wettinia augusta* (Andean piedmont).

5.4.5.3 Small Palms

Erect, single-stemmed: *Asterogyne guianensis* – **erect, multistemmed:** *Bactris monticola* (West), *B. trailiana*, *Catoblastus drudei*, *Geonoma baculifera*, *G. oldemanii*, *Hyospathe elegans* – **climbing, multistemmed:** *Desmoncus* spp. – **acaulescent:** *Geonoma acaulis*, *G. macrostachys*.

5.4.6 Forests on Dry, White-Sandy Soils

5.4.6.1 Medium-Sized Palms

Multistemmed: *Mauritiella aculeata*, *M. peruviana*.

5.4.6.2 Small Palms

Erect, multistemmed: *Bactris arenaria*, *B. simplicifrons*, **climbing, multistemmed:** *Desmoncus* sp.

5.4.7 Forests on Waterlogged, White-Sandy Soils

5.4.7.1 Arborescent Palms

Large – erect, single-stemmed: *Jessenia bataua* subsp. *bataua*, *Mauritia carana*.
Slender – multistemmed: *Mauritiella aculeata*, *M. peruviana*.

5.4.7.2 Medium-Sized Palms

Large – creeping, single stemmed: *Elaeis oleifera*.
Slender – multistemmed: *Euterpe catinga*.

5.4.7.3 Small Palms

Erect, single-stemmed: *Pholidostachys synanthera* – **erect, multistemmed:** *Bactris simplicifrons*, *Lepidocaryum tessmannii* – **climbing, multistemmed:** *Desmoncus* spp.

5.4.7.4 Subterranean-Stemmed, Large-Leafed Palms

Astrocaryum acaule, *Orbignya spectabilis*, *Scheelea insignis*

5.4.8 Submontane and Montane Forests

5.4.8.1 Arborescent Palms

Large – single-stemmed: *Dictyocaryum lamarckianum* (at high elevation); *Iriartea deltoidea*.
Slender – single-stemmed: *Euterpe precatoria* – **multistemmed:** *Prestoea* spp.

5.4.8.2 Medium-Sized Palms

Slender – single-stemmed: *Geonoma helminthoclada*, *Wettinia maynensis*; **multistemmed:** *Oenocarpus mapora*.

5.4.8.3 Small Palms

Erect, single-stemmed: *Geonoma triglochin* – **erect, multistemmed:** *Aiphanes* spp., *Chamaedorea* spp., *Geonoma euspatha*, *G. jussieuana*.

5.4.9 Savannahs

5.4.9.1 Arborescent Palms

Large – single-stemmed: *Acrocomia lasiospatha*, *Mauritia flexuosa*, *Syagrus sancona*; *Copernicia alba* (southern limit of the basin).
Slender – multistemmed: *Astrocaryum vulgare*.

5.4.9.2 Medium-Sized Palms

Slender – multistemmed: *Mauritiella aculeata*, *M. peruviana*; *Syagrus stratincola* (endemic to low dry forest on inselbergs).

5.4.9.3 Small Palms

Erect, multistemmed: *Bactris arenaria*, *B. campestris* – **climbing, multistemmed:** *Desmoncus* spp.

6 Palms as Functional Components in Forest Ecosystems

6.1 Introduction

Palms are an important structural component of forest ecosystems. They dominate the understory of most terra firme forests, as well as the canopy of swamp forests. Their role in the functioning of these is obviously not negligible. However, few data have been published on this topic. Three aspects of particular importance in forest ecosystem functioning will be considered: leaf and fruit productivity, relationship with animals, and some structural features with regard to life forms.

6.2 Palm Productivity

6.2.1 Leaves

6.2.1.1 Arborescent Palms

Data on leaf productivity of *Mauritia flexuosa*, *Euterpe precatoria*, and *Jessenia bataua* subsp. *bataua* are drawn from 2 to 4 years of study carried out from 1985 to 1989 in the lower Ucayali River valley in Peruvian Amazonia by F. Kahn and K. Mejia. Methods and results are first presented and compared with data obtained from published sources on *Euterpe oleracea*, *Jessenia bataua* subsp. *oligocarpa*, *Orbignya phalerata*, *Oenocarpus bacaba*, and *Socratea exorrhiza*. Most of these species offer an economic potential (see Chap. 7).

Methods. Leaf production was measured every 3 months for adult plants, and every 6 months for juveniles. In order to compare the production during the rainy and dry seasons, both adults and juveniles were surveyed every May–June and November–December, i.e., at the transition between rainy and dry seasons. For adult palms of *Mauritia flexuosa*, and juveniles of all three species, all the leaves of the crown were initially marked with paint. For adults of *Euterpe precatoria* and *Jessenia bataua*, the first internode of

Table 50. Leaf production of adults of *Mauritia flexuosa* over a period of 3 years (from August 1985) in Peruvian Amazonia

Year	1985		1986			1987			1988			
Month	08	11	05	08	11	05	08	11	05	08		
Total/year				1st			2nd				3rd	Total
Total/season[a]		D	R		D	R		D	R	D		(3 years)
		(1)	(5)	8	(4)	(3)	10	(6)	–	–	–	Dead
		(1)	(5)	8	(4)	(4)	12	(5)	–	–	–	Dead
		(1)	(5)	8	(4)	(4)	5	(3)	(3)	(1)	8	21
		(1)	(5)	7	(4)	(5)	7	(3)	(4)	(2)	10	24
		(1)	(4)	8	(5)	(3)	6	(3)	(4)	(2)	8	22
		(1)	(3)	7	(4)	(4)	5	(3)	(4)	(1)	8	20
		(1)	(5)	8	(4)	(4)	5	(3)	(3)	(1)	8	21
		(2)	(3)	7	(4)	(5)	6	–	–	–	–	Dead
		(2)	(3)	7	(4)	–	–	–	–	–	–	Dead
		(1)	(4)	7	(4)	(4)	5	–	–	–	–	Dead
		(1)	(4)	7	(4)	(4)	6	(3)	(5)	(1)	9	22
		(1)	(5)	9	(4)	(4)	6	(4)	(4)	(2)	9	24
Mean				7.6			6.6				8.6	22
s				0.7			2.3				0.8	1.5

[a] Production during dry and rainy seasons is given in parentheses; D: dry season; R: rainy season; s: standard deviation.

the trunk, located immediately under the sheath base of the eldest leaf, was painted. This method is actually a measure of abscission, not production of leaves. It was verified that the number of leaves in the crown was constant for the 3–4 years of study in most cases; then abscission corresponds to leaf production. It is also true that the number of leaves in the crown may vary significantly during the adult lifetime of a palm, or over a short period in the case of branch fall or disease. When a variation of the number of leaves in the crown was observed, production was calculated as follows: $P = A + Ne - Nb$ (P: production, A: abscission, Ne and Nb: number of leaves in the crown at the end and at the beginning of the study, respectively). New unmarked leaves and internodes were counted at each survey, and a new mark was painted every 6 months. A large sample of juvenile individuals from 1 m to about 6 m in total height, i.e., the limit of petiole accessibility for painting, were marked. This height of 6 m also corresponds to the limit between acaulescent and trunked juveniles in *Mauritia flexuosa* and *Jessenia bataua*.

Mauritia flexuosa. Twelve arborescent adult palms, six of each sex, were selected, all of them growing in permanently flooded depressions. Five of the female plants, located near the village, were cut down before the study ended by the inhabitants, who collected the fruits. An average of 6.6 to 8.6

Table 51. Leaf production of *Mauritia flexuosa* – acaulescent, juvenile palms – in a permanent swamp forest over a period of 4 years (from June 1985) in Peruvian Amazonia

Year	1985	1986	1986	1987	1987	1988	1988	1989
Month	12	06	12	06	12	06	12	06
6-month results								
No. of leaves	192	286	240	235	302	248	211	175
No. of palms producing	190	258	232	234	277	242	203	175
Mean of leaves/palm producing	1.01	1.10	1.03	1.01	1.09	1.02	1.04	1.00
No. of palms observed[a]	300	300	300	300	300	300	236	236
Mean of leaves/palm observed	0.64	0.95	0.80	0.78	1.01	0.83	0.89	0.74
Annual results								
No. of palms producing		298		300		300		235
No. of leaves produced		478		475		550		386
Mean of leaves/palm producing		1.60		1.58		1.83		1.64
No. of palms producing 3 leaves/year		18		9		19		5
No. of palms producing 4 leaves/year		0		0		1		0

[a] 321 of the 415 acaulescent juveniles were observed initially; 8, 6, and 7 died during the 1st, 2nd, and 3rd year, respectively; data from these were not taken into account. In the 4th year, the plot was incompletely surveyed.

leaves were produced per palm per year; all palms observed produced leaves over the 3 years' period, and the individual production ranged from 5 to 12 leaves per year (Table 50). Mean leaf productions (calculated from 2-year data, Nov. 1985 to Nov. 1987) during the rainy period (Nov–May: 4.1 leaves/palm, s = 0.8) and during the dry period (May–Nov: 3.9 leaves/palm, s = 0.8) was not significantly different ($t = 0.61$, $t^{0.05} = 2.02$ – t: Student–Fisher's test).

Leaf production was measured for acaulescent juveniles in two plots of 1 ha each in a permanent and a seasonal swamp forest (both described in Chap. 3, see Tables 31 and 36). The results will be compared with juvenile palms located on waterlogged soils in a completely open area.

In the permanently flooded swamp forest, 300 palms were sampled during the first 3 years and 236 during the 4th year (Table 51). Mean annual production varied from 1.6 to 1.8 leaves per palm. Mean leaf production during the rainy season (Dec.–June: 1.03 leaves/producing palm, s = 0.05; 0.83 leaf/palm observed, s = 0.16) and during the dry season (June–Dec.: 1.01 leaves/producing palm, s = 0.08; 0.84 leaf/palm observed, s = 0.16) was very similar ($t = 0.25$ and 0.12; $t^{0.05} = 2.45$). During the former the vegetation was flooded by the river and most juveniles were submersed; during the latter the vegetation was flooded by a permanent, acidic water in the depressions which submersed only the roots.

In the seasonal swamp forest, the sample size was 159 palms during the first 3 years and 132 the 4th year (Table 52). Mean annual production was slightly higher than one leaf per producing palm, or from the whole sample.

Table 52. Leaf production of *Mauritia flexuosa* – acaulescent, juvenile palms – in a seasonal swamp forest over a period of 4 years (from May 1985) in Peruvian Amazonia

Year	1985	1986	1986	1987	1987	1988	1988	1989
Month	11	05	11	05	11	05	11	05
6-month results								
No. of leaves	83	133	112	104	109	91	95	75
No. of palms producing	82	119	111	101	102	91	95	75
Mean of leaves/palm producing	1.01	1.12	1.01	1.03	1.07	1.00	1.00	1.00
No. of palms observed[a]	159	159	159	159	159	159	132	132
Mean of leaves/palm observed	0.52	0.84	0.70	0.65	0.69	0.57	0.72	0.57
Annual results								
No. of palms producing		148		155		154		131
No. of leaves produced		216		216		200		170
Mean of leaves/palm producing		1.46		1.39		1.30		1.30
No. of palms producing 3 leaves/year		0		3		3		0

[a] All of the 177 acaulescent juveniles (1–6 m in height) counted in the 1-ha plot were observed initially; 7, 5, and 6 died during the 1st, 2nd, and 3rd year, respectively; data from these were not taken into account. 27 palms were not found during the last survey.

Mean leaf productions during the dry season (May–Nov.: 1.02 leaves/producing palm, s = 0.03; 0.66 leaf/palm observed, s = 0.09) and the rainy season (Nov.–May: 1.04 leaves/producing palm, s = 0.06; 0.66 leaf/per palm observed, s = 0.12) was not significantly different (t = 0.69; $t^{0.05}$ = 2.45).

In an open area on waterlogged soils, only six acaulescent juveniles were observed for 2 years. Annual production was 5.7 (s = 1.2) leaves per palm in 1987 and 6.8 (s = 1.3) in 1988.

Euterpe precatoria. Leaf production was measured for 24 adult palms in a 12-year-old experimental plantation during 3 years, and for 12 palms in a seasonal swamp forest during 2.5 years. In both cases, soils are waterlogged and irregularly flooded after rainfall. Palms are taller, up to 25 m in height, with smaller diameter (mean: 15 cm; s = 0.8 cm) in the seasonal swamp forest than in the plantation (height: <15 m; mean diameter: 17.2 cm, s = 2.2 cm). In both cases, however, the crowns receive full sunlight. In the plantation, average annual production ranged from 9.8 to 11.4 leaves per palm. A mean of 32.1 leaves per palm were produced during the three years (Table 53). Mean seasonal leaf production (calculated from 3-year data, including three rainy seasons, two complete dry, and two half-dry seasons, i.e., Aug.–Nov. 1985 and May–Aug. 1988, lumped together and considered as one) was significantly higher (t = 2.73, $t^{0.01}$ = 2.35) during the rainy (Nov.–May: 5.6 leaves/palm, s = 1.1) than during the dry periods (May–Nov.: 5.1 leaves/palm, s = 1.1). In the seasonal swamp forest, the average annual production was 9.7 and 10.8 leaves per adult palm for the 2 first

Table 53. Leaf production of *Euterpe precatoria* adults in an experimental plantation over a period of 3 years (from August 1985) in Peruvian Amazonia

Year	1985		1986			1987			1988			
Month	08	11	05	08	11	05	08	11	05	08		
Total/year				1st			2nd				3rd	Total
Total/season[a]		D	R		D	R		D	R	D		(3 years)
		(3)	(7)	12	(6)	(5)	11	(5)	(5)	(2)	10	33
		(2)	(8)	14	(6)	(6)	11	(5)	(6)	(2)	10	35
		(3)	(6)	12	(6)	(6)	14	(6)	(6)	(3)	10	36
		(3)	(6)	11	(6)	(6)	13	(6)	(5)	(3)	11	35
		(3)	(5)	9	(5)	(5)	10	(5)	(3)	(2)	9	28
		(3)	(7)	12	(6)	(5)	7	(1)	(5)	(2)	10	29
		(2)	(5)	10	(5)	(4)	9	(6)	(5)	(1)	9	28
		(2)	(7)	14	(7)	(4)	10	(5)	(6)	(3)	10	34
		(2)	(5)	10	(5)	(4)	10	(6)	(5)	(1)	8	28
		(3)	(7)	13	(7)	(7)	11	(5)	(5)	(2)	12	36
		(3)	(7)	13	(6)	(6)	11	(5)	(5)	(3)	11	35
		(2)	(6)	9	(3)	(5)	9	(4)	(5)	(2)	9	27
		(3)	(6)	10	(4)	(5)	10	(5)	(4)	(2)	9	29
		(2)	(7)	12	(6)	(5)	13	(6)	(6)	(3)	10	35
		(2)	(6)	9	(4)	(5)	10	(4)	(5)	(2)	9	28
		(3)	(8)	12	(7)	(9)	14	(4)	(5)	(3)	13	39
		(3)	(6)	10	(6)	(6)	14	(5)	(6)	(3)	11	35
		(3)	(6)	11	(6)	(5)	13	(6)	(4)	(3)	9	33
		(2)	(5)	11	(6)	(6)	10	(5)	(4)	(1)	8	29
		(2)	(5)	11	(5)	(5)	9	(6)	(4)	(2)	9	29
		(2)	(6)	13	(5)	(5)	8	(6)	(4)	(2)	9	30
		(3)	(7)	15	(7)	(7)	12	(5)	(5)	(2)	9	36
		(2)	(6)	9	(4)	(5)	11	(6)	(4)	(2)	9	29
		(3)	(6)	12	(4)	(7)	11	(4)	(7)	(3)	11	34
Mean				11.4			10.9				9.8	32.1
s				1.8			1.6				1.3	3.6

[a] Production during dry and rainy seasons is given in parentheses; D: dry season; R: rainy season; s: standard deviation.

years (Table 54). During the first 6 months of the 3rd year a mean of 5.3 leaves per palm were produced. Measurement for the other half of the year was not completed because of the danger of climbing and marking the highest internode of these high, slender palms. Mean seasonal leaf production (calculated from 2-year data, Nov. 1985 to Nov. 1987) was not significantly higher ($t = 1.72$, $t^{0.05} = 2.03$) during the rainy season (Nov–May: 5.4 leaves/palm, $s = 0.9$) than during the dry period (May–Nov.: 5.0 leaves/palm, $s = 0.7$).

In the same plot of seasonal swamp forest, 131 juvenile palms were observed during the first 3 years, and 108 during the 4th year (Table 55). Average annual production was slightly higher than one leaf per palm when

Table 54. Leaf production of adults of *Euterpe precatoria* in a seasonal swamp forest over a period of 2.5 years (from August 1985) in Peruvian Amazonia

Year	1985		1986			1987			1988	
Month	08	11	05	08	11	05	08	11	02	
Total/year				1st			2nd			Total
Total/season[a]		D	R		D	R		D	R	(2½ years)
		(2)	(5)	8	(5)	(4)	9	(6)	(1)	23
		(3)	(6)	11	(5)	(5)	12	(6)	(3)	28
		(1)	(5)	8	(5)	(5)	9	(4)	(2)	22
		(2)	(5)	10	(5)	(5)	10	(5)	(3)	25
		(3)	(6)	14	(5)	(6)	8	(5)	(3)	28
		(2)	(7)	10	(6)	(4)	11	(5)	(2)	26
		(2)	(7)	12	(6)	(4)	11	(5)	(3)	27
		(3)	(7)	12	(6)	(4)	9	(4)	(3)	27
		(2)	(6)	12	(4)	(5)	8	(5)	(3)	25
		(2)	(6)	8	(4)	(5)	10	(4)	(2)	23
		(2)	(6)	11	(6)	(5)	10	(5)	(3)	27
		(2)	(6)	13	(5)	(6)	9	(5)	(3)	27
Mean				10.8			9.7			25.7
s				2.0			1.2			2.0

[a] Production during dry and rainy seasons is given in parentheses; D: dry season; R: rainy season; s: standard deviation.

Table 55. Leaf production of *Euterpe precatoria* – acaulescent and juvenile stemmed palms – in a seasonal swamp forest, over a period of 4 years (from May 1985) in Peruvian Amazonia

Year Month	1985 11	1986 05	1986 11	1987 05	1987 11	1988 05	1988 11	1989 05
6-month results								
No. of leaves	63	93	73	65	81	62	67	59
No. of palms producing	63	77	72	65	78	62	67	59
Mean of leaves/palm producing	1.00	1.21	1.01	1.00	1.04	1.00	1.00	1.00
No. of palms observed[a]	131	131	131	131	131	131	108	108
Mean of leaves/palm observed	0.48	0.71	0.56	0.50	0.62	0.47	0.62	0.55
Annual results								
No. of palms producing		119		118		120		98
No. of leaves produced		156		138		143		126
Mean of leaves/palm producing		1.31		1.17		1.19		1.29
No. of palms producing 3 leaves/year		4		0		3		0

[a] All of the 153 acaulescent and stemmed juveniles (1–6 m in height) counted in the 1 ha plot were observed initially; 7, 9, and 6 died during the 1st, 2nd, and 3rd year, respectively; data from these were not taken into account. 23 palms were not found during the last survey.

Table 56. Leaf production of juveniles of *Euterpe precatoria* in an experimental plantation over a period of 2 years (from August 1985) in Peruvian Amazonia

DBH (cm)	1986 08	1987 08	Total (2 years)
A) Palms with 10 < DBH < 12 cm			
10.4	5	7	12
11.5	9	10	19
12.2	8	9	17
10.9	6	9	15
11.4	7	9	16
11.5	6	7	13
11.6	6	11	17
11.1	6	12	18
Mean	6.6	9.3	15.9
s	1.3	1.8	2.4
B) Palms with DBH < 10 cm			
9.8	10	7	17
7.4	3	5	8
5.9	6	5	11
7.5	4	5	9
6.1	3	6	9
5.1	6	2	8
9.3	5	8	13
Mean	5.3	5.4	10.7
s	1.4	1.9	3.3

s: standard deviation.

calculated from the productive individuals, or from the whole sample; the maximum production of three leaves per year was reached by only four and three individuals during the 1st and 3rd year, respectively. Mean leaf productions during the dry season (May–Nov.: 1.01 leaves/producing palm, s = 0.02; 0.57 leaf/palm observed, s = 0.07) and during the rainy season (Nov.–May: 1.05 leaves/producing palm, s = 0.11; 0.56 leaf/palm observed, s = 0.11) was not significantly different (t = 0.87 and 0.19; $t^{0.05}$ = 2.45). In a permanent swamp forest where *E. precatoria* density is low, only five of the eight individuals initially studied were still alive after 3 years. Mean annual production ranged from 1.4 to 2.6 leaves per palm. In the plantation, leaf production of 15 palms was measured during 2 years. These palms, which were never observed flowering before, present a trunk diameter smaller than 12 cm. Eight of them (10–12 cm diameter) produced a mean of 6.6 and 9.3 leaves during the 1st and 2nd year, respectively (Table 56A); the other seven (diameter less than 10 cm) produced an annual mean of 5.3 to 5.4 leaves (Table 56B). The correlation calculated between the diameter and number of leaves produced during the 2 years (n = 15, r = 0.849) clearly suggests that leaf productivity increases with the diameter of juvenile palms.

Table 57. Leaf production of *Jessenia bataua* – acaulescent, juvenile palms – in a seasonal swamp forest, over a period of 4 years (from May 1985) in Peruvian Amazonia

Year Month	1985 11	1986 05	1986 11	1987 05	1987 11	1988 05	1988 11	1989 05
6-month results								
No. of leaves	60	112	79	80	92	85	68	93
No. of palms producing	60	111	79	80	92	85	68	93
Mean of leaves/palm producing	1.00	1.01	1.00	1.00	1.00	1.00	1.00	1.00
No. of palms observed[a]	238	238	238	238	238	238	222	222
Mean of leaves/palm observed	0.25	0.47	0.33	0.34	0.39	0.36	0.31	0.42
Annual results								
No. of palms producing		163		148		175		152
No. of leaves produced		172		159		177		161
Mean of leaves/palm producing		1.06		1.07		1.01		1.06
No. of palms producing 2 leaves/year		9		11		2		8

[a] All of the 258 acaulescent juveniles (1–6 m in height) counted in the 1-ha plot were observed initially; 6, 7, and 7 died during the 1st, 2nd, and 3rd year, respectively; data from these were not taken into account. 16 palms were not found during the last survey.

***Jessenia bataua* subsp. *bataua*.** Leaf production of 20 adult palms was measured in a seasonal swamp forest during 3 years. Fifteen of them produced one to four leaves during the 1st year, with a mean of 1.9 (s = 0.8) leaves per palm. Nineteen of the 20 palms had produced at least one leaf by the end of the 3rd year, with a mean of 5.3 (s = 2.6) leaves produced per palm in 3 years. One palm, still alive, produced no leaf during the same period. The variability of individual growth rate is very high; from one to ten leaves were produced during the 3 years (CV = 96.6%).

In the same plot, the sample was of 238 acaulescent juvenile palms for the first 3 years, and of 222 for the 4th year (Table 57). Annual mean production was one leaf per palm when calculated only from the productive individuals, and lower than one leaf per palm when calculated from the whole sample. Annual production of two leaves was the maximum reached by acaulescent juveniles of this species. During the dry and rainy seasons, leaf production does not vary when calculated only from producing palms (May–Nov.: 1.00 leaf/palm, s = 0.00; Nov.–May: 1.00 leaf/palm, s = 0.01), and is not significantly different when calculated from the whole sample (May–Nov.: 0.32 leaf/palm, s = 0.06; Nov.–May: 0.40 leaf/palm, s = 0.06; t: 2.26, $t^{0.05}$ = 2.45).

In Ecuadorean Amazonia, Borgtoft Pederson and Balslev (1990) calculated annual productions of 2.0–6.9 leaves from eight trunked individuals observed during 318–373 days.

***Jessenia bataua* subsp. *oligocarpa*.** Sist (1989c) measured an average annual production of three leaves for adult palms (11 individuals observed during 20 months). Leaf production of seedlings and acaulescent juveniles is lower with 0.76 and 0.81, produced per year leaf respectively.

***Orbignya phalerata*.** Anderson (1983) concluded that leaf production in babassu palm peaks during the rainy reason. For a 1-year period the production is very similar in three different climatic regions (means of 3.9, 4.1, and 3.7 were found in wet, intermediate, and dry regions, respectively), all located in the Brazilian state of Maranhão at the southeastern limit with the Amazon basin. Leaf production is, however, significantly different between these ecosystems within the intermediate region with 3.7, 4.4, and 5.1 leaves per year in secondary, primary forest, and pasture, respectively. The author related the lower productivity in the secondary forest to shading which was highest because "the dense, uniform babassu stands formed a closed, interlocking canopy" (p. 49).

***Euterpe oleracea, Oenocarpus bacaba*, and *Socratea exorrhiza*.** In the Mapane region (Surinam), Van der Steege (1983) counted the leaves produced by these species during 1 year, considering all palms with a height over 1.5 m in six 0.25-ha plots. For *E. oleracea* and *O. bacaba*, an average production of one leaf per palm was calculated and average leaf lifetime was estimated at 3.5 and 4.0 years, respectively, from 16 and 77 individuals. For *S. exorrhiza*, an average annual production of 1.5 leaves per palm from six individuals was calculated. Leaf lifetime was estimated at 3.5 years. For each species, results are lumped together for all height classes and cannot be significantly compared with those of the former species.

6.2.1.2 Understory Palms

Sist (1989c) studied two understory species in French Guiana, *Astrocaryum sciophilum*, a medium-sized palm with a trunk diameter of around 15 cm, and *A. paramaca*, a subterranean-stemmed palm with large leaves. Adult palms of *A. sciophilum* produce a mean of 1.07 leaves per year (14 individuals observed during 22 months); growth rate is slightly higher in seedlings (1.5 leaves/year), and lower in the first juvenile phase (0.61 leaf/year). Leaf production of the population studied is higher during rainy periods. Adults of *A. paramaca* produce a mean of 1.19 leaves per year (20 individuals observed during 21 months); as in the former case, leaf production is higher in seedlings (2.38 leaves per year), and decreases in the first juvenile phase (0.89 leaf per year). Van der Steege (1983) also provided data on leaf production of *Astrocaryum sciophilum* and *A. paramaca*, and of *Geonoma baculifera* from Surinam. A total of 65, 282, and 44 palms with a height over 1.5 m were observed for 1 year, respectively; data are lumped together for all height classes. Annual mean production is 1.1 leaves for *A. sciophilum*, and 1.3 leaves for *A. paramaca*. Estimate of leaf lifetime is 8.5 years for the former, and 5.6 years for the latter. Mean production of *G. baculifera* is 2.0 leaves; estimate of leaf lifetime is 3.2 years.

6.2.1.3 Discussion

Adults of *Mauritia flexuosa* and *Euterpe precatoria* (with 6.6 to 8.6 and 9.7 to 10.8 leaves/palm/year, respectively) are more productive than *Orbignya phalerata* (3.7 to 5.1 leaves/palm/year), and than both subspecies of *Jessenia bataua* (<2 and 3 leaves/palm/year). These latter differ from each other by the variability of interindividual growth rate, which is very low in *J. bataua* subsp. *oligocarpa* and high in *J. bataua* subsp. *bataua*. However, Borgtoft Pederson and Balslev's data (1990) indicate that Ecuadorean populations can present higher growth rates.

Leaf production of *Orbignya phalerata* is higher during the rainy season in the southeastern region adjacent to the Amazon basin (Maranhão State, Brazil) where the dry season is well marked. On the other hand, in western Amazonia near Iquitos, where the dry season is slightly marked, leaf production of adults of *Mauritia flexuosa* is remarkably similar during the rainy and dry seasons. Seasonal difference of leaf production is significant only with adults of *Euterpe precatoria* growing in a monospecific plantation on waterlogged soils. In the seasonal swamp forest, the seasonal difference of leaf production measured in a smaller sample (12 palms with 2-year data vs.

Table 58. Reproductive efficiency in a *Jessenia bataua* population located in Peruvian Amazonia

	No. of flowers[a]	No. of fruits[a]	Reproductive efficiency (%)
	9599	1959	20.4
	3938	528	13.4
	17127	2662	15.5
	9403	906	9.6
	9737	579	6.0
	11323	2479	21.9
	13524	3318	24.5
	7124	1452	20.4
	11730	2615	22.3
	3645	450	12.3
	5597	595	10.6
	15141	3222	21.3
Mean	9824.0	1730.4	16.5
s	4256.5	1106.4	6.0

[a] Calculated from 30 rachillae counted on each inflorescence or infructescence. s: standard deviation.

24 palms with 3-year data in the former case) is not significantly higher during the rainy season. Leaf production of adults of *Jessenia bataua* is too low to be analyzed in relation to seasons on the 3 years' study.

Leaf production of juvenile palms is higher in open areas, as data for *M. flexuosa* demonstrate, or in a plantation, the canopy of which is more open than that of a seasonal swamp forest, as pointed out by *E. precatoria* data. In swamp forests in western Amazonia, seasonal production means of juveniles of *M. flexuosa*, *E. precatoria*, and *J. bataua* are not significantly different. This is particularly surprising in the case of juveniles of *M. flexuosa* located in the permanently flooded depressions; their leaves are submersed by river water during 1–3 months at the time of highest flooding during the rainy season.

Within a species, the adults produce more leaves than the juveniles. This is true of arborescent as well as understory species. However, the growth rate of seedlings of understory palms (*Astrocaryum paramaca* and *A. sciophilum*) is higher than in adult plants, a phenomenon not observed for the arborescent *Jessenia bataua* subsp. *oligocarpa* (Sist 1989c).

6.2.2 Flowers and Fruits

There is a lack of data on palm phenology in Amazonian forests. Most information is limited to economically important species.

6.2.2.1 Arborescent Species

Euterpe precatoria. In a Peruvian population, palms produced two to six inflorescences per year. Flowering at the beginning of the dry season (May), the fruits are ripe in December-January. Maturation time is 7–8 months. A mean of 6039 (s = 396; n = 5) fruits per infructescence was counted.

***Jessenia bataua* subsp. *bataua*.** In a Peruvian population a mean of 9824 flowers and 1730 fruits per bunch were calculated counting flower scars and fruits on 30 rachillae of each of 12 infructescences from eight palms. The reproductive efficiency (number of fruits/number of flowers × 100) varies from 6.0 to 24.5% in this population (Table 58), and from 3.4 to 14.5% in a Colombian population (Balick 1986). In Ecuador, Borgtoft Pederson and Balslev (1990) calculated an annual production of 2 to 4.6 inflorescence buds per palm.

***Jessenia bataua* subsp. *oligocarpa*.** The phenology of *Jessenia bataua* subsp. *oligocarpa* was studied in French Guiana by Sist (1989c) for 2 years (1986–1987). Of the 30 adults observed, 28 (93.3%) and 24 (80.0%) were fertile in

1986 and 1987, respectively. Fruits were produced from October 1986 to March 1987 with a peak in November. Counting the inflorescences and leaf scars on the trunk, the author pointed out the alternance of vegetative phases, which last 16 to 22 months, with reproductive phases every other year. The fruit maturation phase is about 16 months and the mean number of fruits per infructescence is 1128 (from 24 infructescences from 18 palms); one palm produced 4967 fruits on three infructescences. The mean abortive flower rate was 86.2%.

Mauritia flexuosa. In Peruvian Amazonia, each female tree produces two to six infructescences yearly, with a mean of 493 (s = 34.7) fruits calculated from 10 infructescences; one palm was highly productive with a bunch bearing 2190 fruits. From ten 0.5-ha plots, Salazar and Roessl (1977) calculated a mean of 2.6 infructescences per palm, 131 infructescences per ha bearing a total of 75 291 fruits, i.e., a mean of 577 fruits per infructescence. The fruiting period extends from June to November, with peak from July to September, during the dry season; fruits are available in Iquitos markets during all months except December, January and February (Padoch 1988; Vásquez and Gentry 1989). In Colombian Amazonia (Araracuara, 0°37′S; 72°20′W), the flowering period extends from September to October, and the fruiting period from May to mid-July; a mean production of four

Table 59. Fruiting periods of several palm species

	Months											
	01	02	03	04	05	06	07	08	09	10	11	12
Acrocomia lasiospatha[a]							———	———	———	———	———	– –
Astrocaryum aculeatum[a]	–	———	———	———	—	– –						
Astrocaryum jauari[c]	–	———	———	–								
Astrocaryum vulgare[a]	–	———	———	———	———	———	—	– –	– –			
Bactris maraja[a]	–	———	———	———	———	–						
Euterpe oleracea[a]	– –	–	–	– –	–	– –	–	———	———	———	———	———
Euterpe precatoria[d]	———	———	–	– –	–				–	–	———	———
Jessenia bataua[a,d,e]	———	———	–	– –	–					———	———	———
Mauritia flexuosa[a,d]					–	–	———	———	———	–	–	
Mauritiella aculeata[a]	———	———	—	–	– –	–						
Oenocarpus bacaba[a]	–	———	———	———	–	–				–	– –	–
Oenocarpus distichus[a]	———	———	———	—	– –					———	———	———
Oenocarpus mapora[a,d]	———	———	—	–	– –						———	———
Oenocarpus minor[a]	———	———	–	–								
Orbignya phalerata[b]	–									–	———	———

[a] Cavalcante (1974) from eastern and central Amazonia.
[b] Anderson et al. (1988) from the limitrophe southeastern Maranhão State.
[c] Piedade (1984) from central Amazonia.
[d] Kahn F., pers. observ., from western Amazonia.
[e] *Jessenia bataua* subsp. *bataua*.

infructescences (CV: 39.7%) per palm per year, with an average of 470 fruits (CV: 51.8%) per infructescence was reported by Urrego Giraldo (1987).

Orbignya phalerata. Flowering mainly occurs during the rainy season, peaking from January to March. Fruiting peaks in October–December; fruit maturation time is an average of 9 months. An analysis of ten infructescences from palms growing in a secondary forest shows that 78.4% of flowers became fruits, 19.9% were aborted or not fertilized, and 1.7% were attacked by insect larvae; analysis of inflorescences of palms from pasture produced similar results. Parthenogenesis does not occur in babassu palm (Anderson 1983; Anderson et al. 1988).

Other Arborescent Species. In central Amazonia (Anavilhanas Archipelago on the Rio Negro), *Astrocaryum jauari* produces an average of 106 fruits per infructescence (29 to 218, from ten infructescences); flowering occurs during August–September at the beginning of the river subsidence, and fruiting peaks during May–June at the highest water level; fruit maturation time is about 9 months (Piedade 1984). Cavalcante's (1974) fruiting calendar for edible fruit species of Amazonia is among the few phenology data available. Several palm species are considered (Table 59).

6.2.2.2 Understory Species

Data on phenology of six understory species, *Astrocaryum paramaca*, *A. sciophilum*, *Bactris acanthocarpoides*, *B. rhaphidacantha*, *Geonoma oldemanii*, and *Scheelea* spec. nov. are reported by Sist (1989c) for the period 1986 and 1987 (Table 60). Flowering and fruiting phases are: annual, discontinuous, and irregular for *A. sciophilum*, regular in the other cases; flowering is after a dry period (*Geonoma oldemanii*), at the beginning of the rainy season (*A. paramaca*, *B. acanthocarpoides*, *Scheelea* spec. nov.), and after the rainy season (*B. rhaphidacantha*). Mean fruit production per infructescence is 65 (from 20 infructescences from 13 palms) in *A. sciophilum*, and 93 (from 44 infructescences from 33 palms) in *A. paramaca*. Mean abortive rates are 35.9 and 19.8%, respectively. Parthenocarpic fruits were not observed. Fruit maturation periods vary according to the species: 4–6 months (*A. sciophilum*), 5 months (*A. paramaca*), 3 months (*B. acanthocarpoides*), 10 months (*B. rhaphidacantha*), and 9 months (*G. oldemanii*). The difficulty of embarking upon such a study is pointed out by the case of *Scheelea* spec. nov., a monoecious, subterranean-stemmed species with staminate and pistillate flowers on separate inflorescences. The 60 adult palms observed produced only two female inflorescences, neither of which produced ripe fruit; one was abortive early and the other developed green fruits which were eaten by a rodent.

Table 60. Phenology of understory palms in French Guiana. (After Sist, 1989c)

	Sample	No. of palms fertile (%)	Fertility rate[b] (%)	Fl [month(s)]	Fr
Astrocaryum paramaca	40	18[a] (45.0)	93.8	–	–
1986		18 (45.0)	–	11	03–04
1987		12 (30.0)	–	–	04
Astrocaryum sciophilum	77	40[a] (51.9)	31.7	–	–
1986		39 (50.7)	32.1	–	05–06
1987		13 (16.9)	30.8	–	–
Bactris acanthocarpoides	41	13[a] (31.7)	62.5	–	–
1986		13 (31.7)	–	11	04–06
1987		7 (17.1)	–	12	01–06
Bactris rhaphidacantha	31	17[a] (54.8)	–	–	–
1986		17 (54.8)	–	04	05–06
1987		7 (22.6)	–	–	01–02
Geonoma oldemanii	97	46[a] (47.4)	35.3	–	–
1986		33 (34.0)	–	09–10	05
1987		33 (34.0)	–	09–10	05
Scheelea sp.	60	15[a] (25.0)	0.0	–	–
1986		12 (20.0)	–	11	–
1987		9 (15.0)	–	11	–

[a] Data for the period 1986–1987.
[b] Fertility rate (%): no. of infructescences/total no. of inflorescences. Fl: flowering; Fr: fruiting.

Elaeis oleifera in seasonal swamp forests in Peruvian Amazonia produces 213 to 1508 fruits (mean: 707, calculated from six bunches), of which 11.2 to 63.1% (mean 39.5%) are parthenocarpic (Kahn and Mejia, unpubl. 1986).

6.2.2.3 Discussion

Understory palms produce fewer fruits than arborescent palms, with the exception of *Elaeis oleifera*. This species is actually a large palm with a prostrate trunk, with a crown size similar to that of the African, erect-trunked *E. guineensis* Jacquin, and thus unrepresentative of understory palms.

As shown by Sist's study (1989c), the abortive rate is high for both arborescent and understory palms; many seeds are parasited by insects (see Sect. 6.3.3). Several of the understory species – small, single-stemmed or subterranean–stemmed, or acaulescent palms – form dense populations in spite of producing only a few seeds, and having nothing but seed production and dispersal as the means to maintain or increase the size of their populations. This is the case for *Astrocaryum paramaca*, *A. sociale*, *A. sciophilum*, *Geonoma acaulis*, *G. poeppigiana*, *Orbignya polysticha*, *O. spectabilis*,

Pholidostachys synanthera, etc. In contrast, adult arborescent palms which produce a large number of seeds are found in very low density in the closed-canopied terra firme forests, and in high density in the open-canopied swamp forests (see Chap. 3). Their seedlings and juveniles form dense populations in the understory in all cases, however. The effect of forest architecture and dynamics on palm population dynamics was discussed in Chapter 4.

Flowering and fruiting periods of widely distributed species vary from east to west of Amazonia and depend on local climates. Nevertheless, most terra firme species in eastern Amazonia present a flowering peak at the end of the dry season (Sist 1989c). This is also true of trees (Sabatier 1985). Fruiting usually peaks during the rainy season in most species, with the exception of *Euterpe oleracea* and *Mauritia flexuosa*, which produce fruits the dry seasons (Table 59); both species grow in permanent swamp forests. The period of mature fruit of *Astrocaryum jauari* corresponds to the highest flooding in central Amazonia (Piedade 1984). Seasonal changes in palm phenology related to "El Niño" have been pointed out by De Steven et al. (1987) in Panama.

6.2.3 Productivity and Forest Ecosystems

Arborescent palms in swamp forests are a significant source of litter due to their high density. In Peruvian Amazonia, for instance, on a 1-ha plot in a permanently flooded swamp forest, the *Mauritia flexuosa* population is composed of 138 adults producing a mean of 6.6–8.6 leaves per year, 415 acaulescent juveniles (with height > 1 m) producing a mean of 1.66 leaves per year (Table 51), and 92 trunked juveniles, the individual productivity of which is between 1.6 and 6.6 leaves per year. Assuming a mean of three leaves produced yearly for the third category, and 6.6 leaves (lower mean) for the first, a total of 1876 large leaves per year are produced by *Mauritia flexuosa* on 1 ha in this swamp forest. Morever, each adult palm produces from three to six inflorescences yearly. Annual production including leaves (88.5%) and sexual parts (11.5%) reaches 15.8 t/ha (dry weight, Table 61). Thus, it is not surprising that the soils of these swamps are essentially composed of organic matter, several meters in depth in acidic water (histosol).

In a seasonal swamp forest the mean annual productivities of the three species, *Mauritia flexuosa*, *Jessenia bataua*, and *Euterpe precatoria* are 566.9, 335.6, and 843.8 leaves, respectively (calculated as explained in Table 62). A total of 1746 leaves are produced annually on the 1-ha plot by these three species, that is 5.1 t (dry weight). The total production (including leaves and sexual parts) is 6.0 t/year, of which 69.8, 20.9, and 9.3% are produced by *M. flexuosa*, *J. bataua*, and *E. precatoria*, respectively (Table 63). The proportion represented by the leaves is 90.0, 75.9, and 68.0%, respectively, for the

Table 61. Productivity of *Mauritia flexuosa* on 1-ha swamp forest in Peruvian Amazonia

Leaves (88.5%)		Annual leaf production (No.)	Leaf dry weight[a] (kg)	Annual leaf dry weight (kg)
Adults		910.8	12.3	11202.8
Stemmed juveniles		276.0	9.0	2484.0
Acaulescent juveniles		688.9	0.5	344.5
Total		1875.7		14031.3
Sexual parts (11.5%)	**No. of palms producing**	**No. of infl. /palm /year**	**Dry weight[b] /infl. (kg)**	**Annual dry weight of sexual parts (kg)**
	138	4	3.3	1821.6

[a] Mean from 2 leaves dried for adults, 2 for stemmed juveniles, and 4 for acaulescent juveniles.
[b] Mean from one inflorescence and one infructescence dried. The material was cut into small pieces and dried at 60 °C over a period of 3 weeks.

three species. The waterlogged soil, a humic gleysol, is characterized by a high concentration of organic matter in its upper part; under stands of *M. flexuosa* an organic soil is formed.

The production of dry matter from leaves and sexual parts of a population of *Mauritia flexuosa* in a permanently flooded swamp is very high as suggested by a comparison with the production of falling litter in a terra firme forest of French Guiana – 15.8 t/ha/year in the former vs. 7.8 t/ha/year in the latter (Puig and Delobelle 1988), even though these data are not completely comparable. The production of a mixed population of *Euterpe precatoria*, *Jessenia bataua*, and *Mauritia flexuosa* in a seasonal swamp forest is also high (6.0 t/ha/year) if we consider that (1) the three species represent only 36.3% of the total basal area (DBH > 15 cm), and (2) only leaf and sexual part production is taken into account. Anderson (1983) estimated the annual above-ground dry matter production (including trunk) of mature *Orbignya phalerata* at 4.3, 13.3, and 11.6 t/ha/year in a primary forest, secondary forest and pasture, respectively; leaves and sexual parts represent 3.4, 10.5, and 9.4 t/ha/year (79.0 to 80.9%), and leaves alone, 2.3, 7.4, and 5.9 t/ha/year (50.5 to 55.4%) in each ecosystem, respectively. Leaves represent 51.7 and 39.4% of the lifetime production in *Jessenia bataua* subsp. *bataua* and in *Orbignya phalerata*, respectively (Anderson 1983; Balick and Anderson 1986/1987).

In terra firme forests of Surinam, the total phytomass of leaves of all palms with a height over 1.5 m varied from 294 to 1212 kg/ha (dry weight). Primary production of leaf material estimated from 1 year's study ranges from 47 to 188 kg/ha/year (Van der Steege 1983). In a terra firme central

Table 62. Estimate of annual leaf production of the three species, *Euterpe precatoria*, *Jessenia bataua*, and *Mauritia flexuosa*, in 1 ha of seasonal swamp forest in Peruvian Amazonia

	Euterpe precatoria	*Jessenia bataua*	*Mauritia flexuosa*
No. of adults/ha[a]	45	48	32
Percentage of adults producing leaves	100[b]	75[c]	100[d]
Mean No. of leaves produced by 1 adult/year	10.3[b]	1.9[c]	6.6[e]
Total No. of leaves produced by adults/ha/year (A)	463.5	68.4	211.2
No. of juveniles (1 m < h < 6 m)/ha[a]	153[f]	258[g]	177[h]
Percentage of juveniles producing leaves	90.81[f]	68.17[g]	96.67[h]
Mean No. of leaves produced by 1 juvenile/year	1.24[f]	1.05[g]	1.36[h]
Total No. of leaves produced by juveniles/ha/year (B)	172.3	184.7	232.7
No. of juveniles (6 m < h < 10 m)/ha[a]	52	55	41
No. of leaves produced by 1 juvenile/year[i]	4	1.5	3
Total No. of leaves produced by juveniles/ha/year (C)	208	82.5	123
Total No. of leaves produced by palms (with h > 1 m)/ha/year (A + B + C)	843.8	335.6	566.9

From: [a]Table 31. [b]Table 54. [c]6.2.1. [d]Table 50. [e]6.2.3. [f]Table 55. [g]Table 57. [h]Table 52. [i]Estimate.

forest in central Amazonia, Klinge and Rodrigues (1971) estimated that 17% of leaf biomass consists of palm leaves.

6.3 Palms and Animals in Amazonian Forests

Animals play an important role as pollinators and seed dispersers of palms, and palms provide habitats for many animals, primarily arthropods. Lepesme (1947) published a large inventory of insects related to palms. Only some examples from Amazonia are presented to emphasize the great diversity of entomofauna on palms; the economic impact is pointed out because this fauna is able to damage the African oil palm and coconut tree plantations.

6.3.1 Pollination

Henderson (1986a) completely reviewed the literature on palm pollination and concluded that "anemophily appears uncommon and derived" and that "palms are predominantly entomophilous". He distinguished three basic

Table 63. Production of leaves and sexual parts (dry weight[a]) of three palms, *Euterpe precatoria*, *Jessenia bataua*, and *Mauritia flexuosa*, in 1 ha seasonal swamp forest in Peruvian Amazonia. (Leaf production estimate is explained in Table 62)

Euterpe precatoria

Leaves (68.0%)	Annual leaf production (No.)	Leaf dry weight (kg)	Annual leaf dry weight (kg)
Adults	463.5	0.7	324.5
Trunked juveniles	208	0.2	41.6
Acaulescent juveniles	172.3	0.1	17.2
Total	843.8		383.3

Sexual parts (32.0%)	No. of palms producing	No. of infl./palm /year	Dry weight /infl. (kg)	Annual dry weight of sexual parts (kg)
	45	4	1.0	180.0

Jessenia bataua

Leaves (75.9%)	Annual leaf production (No.)	Leaf dry weight (kg)	Annual leaf dry weight (kg)
Adults	68.4	9.4	643.0
Trunked juveniles	82.5	2.8	231.0
Acaulescent juveniles	184.7	0.5	92.4
Total	335.6		966.4

Sexual parts (24.1%)	No. of palms producing	No. of infl./palm /year	Dry weight /infl. (kg)	Annual dry weight of sexual parts (kg)
	48	2	3.2	307.2

Mauritia flexuosa

Leaves (90.0%)	Annual leaf production (No.)	Leaf dry weight (kg)	Annual leaf dry weight (kg)
Adults	211.2	12.3	2597.8
Trunked juveniles	123	9.0	1107.0
Acaulescent juveniles	232.7	0.5	116.4
Total	566.9		3821.2

Sexual parts (10.0%)	No. of palms producing	No. of infl./palm /year	Dry weight /infl. (kg)	Annual dry weight of sexual parts (kg)
	32	4	3.3	422.4

[a] See Table 61. For *Euterpe precatoria*, the sample was 4 leaves for adults, 4 for medium-sized juveniles, and 6 for smaller juveniles.

patterns: cantharophily (beetle pollination), mellitophily (bee pollination), and myophily (fly pollination). Two groups of beetles closely associated with palms are Curculionidae and Nitidulidae. This is true with *Astrocaryum* cf. *gratum* as well as with *Bactris* spp. and *Desmoncus* cf. *mitis* which, however, develop a more complex pollination strategy: the beetles pollinate the palms and reproduce in the staminate flowers (Listabarth 1992). As concluded by the author, *Bactris* and *Desmoncus* breed their own pollinators. Bees of the genera *Melipona*, *Apis*, and *Trigona*, as well as flies of the subfamilies Calliphoridae, Syrphidae, and Drosophilidae are pollen-collecting visitors on palms; the latter appear to be more common in understory species (Syed 1979; Beach 1984; Henderson 1985; Barfod et al. 1987; Olesen and Balslev 1990). More recently, Anderson et al. (1988) reported that babassu palms (*Orbignya phalerata*) are insect-pollinated when located in secondary forest, but wind-pollinated when located in a pasture, with similar efficiency in both cases (abortive rate: 19.9 and 17.1%, respectively). According to Listabarth (1992), the protandrous *Aiphanes* is mainly anemophilous though entomophily also occurs.

Olesen and Balslev (1990) described a case of Bakerian mimicry in *Geonoma macrostachys* from Ecuadorean Amazonia. This species is monoecious and protandrous, like other geonomoid palms (Wessels Boer 1968; Uhl and Dransfield 1987). The pistillate flowers resemble the staminate in size, color, scent, and form by developing a long staminodial tube; they have no rewards for the pollinators and probably attract insects by mimicking the staminate. Insects of 22 and 10 species visited the inflorescences during the staminate and pistillate phases, respectively. Trigonidae and Drosophilidae are considered as the most important pollinators.

6.3.2 Fruit Dispersal

Due to their high content of lipids, proteins, and glucids, fruits of most neotropical palm species are consumed and dispersed by small vertebrates, mainly mammals (bats, rodents, and monkeys) and birds (Greenhall 1965; Janzen et al. 1976; Heaney and Thorington 1978; Janzen 1978; Smythe 1978; Yeaton 1979; Vandemeer et al. 1979; Snow 1981; Smythe et al. 1983; Rylands 1987; Sist 1989c; Henderson 1990). Zona and Henderson (1989) provide a review of the literature on the topic, emphasizing the role of animals on palm dispersal. They conclude that "both obligate and opportunistic frugivores are capable of dispersing seeds", and "there is little evidence for obligate-plant-animal mutualisms in palm seed dispersal ecology".

Information relating to Amazonian palms only will be presented here.

Mammals. Rodents consume fruits of *Astrocaryum paramaca* (*Sciurus aestuans*, *Proechimys cuvieri*, *P.* sp.), *A. sciophilum* (*S. aestuans*, *Myoprocta* *exilis*), *A. vulgare* (*Dasyprocta agouti, S. aestuans*), *Euterpe oleracea* (*P. cuvieri*), *Jessenia bataua* subsp. *oligocarpa* (*S. aestuans*), *Maximiliana maripa* (*Dasyprocta agouti, S. aestuans*), *Oenocarpus bacaba* (*Proechimys cuvieri*), and *Socratea exorrhiza* (*M. exilis*), as reported in French Guiana by Guillotin (1981), Charles–Dominique et al. (1981), Sabatier (1983), and Sist (1989c). Anderson (1983) reports that the fruit of *Orbignya phalerata* is disseminated by pacas (*Agouti paca*) and agoutis (*Dasyprocta punctata*); the mesocarp is eaten by several animals which have no significant role in dispersal of this palm (porcupines: *Coendu prehensilis, C.* sp.; spiny rats: *Proechimys longicaudatus, Mesomys hispidus*; pecaries: *Tayassu pecari*; *T. tajacu*). The fruit of *Aphandra natalia* is dispersed by *Dasyprocta* sp. (Borgtoft Pederson and Balslev 1990)

Bats consume fruits of *Socratea exorrhiza* (*Artibeus jamaicensis, A. lituratus*), as noted by Huber (1910), and in French Guiana by Charles-Dominique et al. (1981), and de Foresta et al. (1984).

Monkeys consume fruits of *Euterpe oleracea, E. precatoria* (*Ateles paniscus*), *Jessenia bataua* (*Ateles paniscus, Cebus nigrivitatus*), and *Socratea exorrhiza* (many species) as described in Amazonia by Van Roosmalen (1980), Milton (1980), Rylands (1987), and Sist (1989c). *Cebus apella* and *Aotus trikingatus* break the kernels of immature fruits of babassu and drink the liquid endosperm (Anderson 1983), and *Cebus apella* does the same with the nut of *Astrocaryum chambira* (Izawa and Mizuno 1977; Izawa 1979).

Birds. Several species consume fruits of *Astrocaryum gynacanthum, Euterpe precatoria* (*Rupicola rupicola*), *E. oleracea* (*Perissocephalus tricolor, Psophia crepitans, Ramphastos tucanus, Rupicola rupicola*), *Jessenia bataua* (*Amazona farinosa, A. ochrocephala, Penelope marail, Perissocephalus tricolor, Pionites melanocephala, Pionus fuscus, Ramphastos tucanus*), and *Oenocarpus bacaba* (*Penelope marail, Perissocephalus tricolor, Ramphastos tucanus, R. vitellinus, Rupicola rupicola, Selenidera culick, Trogon melanurus*), as reported in French Guiana by Erard et al. (unpubl.), Sist (1989c), and Tostain and Sabatier (1988).

Fishes. Several riparian species are dispersed and eaten by fishes. One of these fishes, the famous *Colossoma macropomum*, called "tambaqui" in Brazil, and "gamitana" in Peru, is reported to eat and disperse the fruits of *Astrocaryum jauari* (Goulding 1980). Several other fishes which eat the fruit of this palm are listed by Piedade (1984): *Colossoma bidens, Mylossoma* sp., *Serrasalmus* spp. (Serrasalmidae), *Brycon melanopterus, B.* sp. (Characidae), *Leporinus* spp. (Anostomidae), *Phractocephalus hemioliopterus, Paulicea lutkeni, Rhamdia schomburgkii* (Pimelodidae), *Lithodoras dorsalis, Megalodoras irwini, Oxydoras niger* (Doradidae), *Semaprochilodus* spp. (Prochilodontidae).

6.3.3 Palms as Habitats

Palms constitute the habitat of many animals, most of which are palm predators, eating fruits or seeds, or stinging and sucking leaf parts. Insects are particularly frequent and diversified in palms; other arthropods and small vertebrates are also found.

Coleoptera. The fruits of *Scheelea rostrata* beneath the parent tree are attacked by bruchids (Janzen 1971). The tribe Pachymerini (Bruchidae: Pachymerinae) is specialized in parasiting seeds of palms (Borowiec 1987). Bondar (1936) reported seed predation of babassu palms by *Pachymerus nucleorum*; Anderson (1983) corroborated the fact and also noted *Caryobruchus lipismatus* as a seed predator. Sist (1989a) reported high rates of seed parasitism by bruchids: *Caryoborus serripes* was found in 50% of the seeds of *Astrocaryum sciophilum* with up to three insects per seed; *Pachymerus* sp. was found in 15 of 80 seeds of *Astrocaryum paramaca*; *Caryoborus* sp. were also encountered in seeds of *Jessenia bataua*, with lower frequency however, just two out of the 60 seeds in incubation. None of the seeds of *Oenocarpus bacaba* in incubation was parasitized. In Peruvian Amazonia, *Pachymerus* sp. was found in seeds of *Astrocaryum* cf. *chonta*, and *Caryoborus* sp. in *Oenocarpus balickii*.

Many species of Curculionidae are palm predators. Larvae of several species of *Derelomus* (Erirhininae, Derelomini) develop in the male inflorescences, on which their adults are in high density, as reported by Bondar (1941). This author (1940, 1942, 1943) also provided a list of such insects with several notes on their biology. All staminate flowers of *Astrocaryum*, *Orbignya*, *Phytelephas*, and bisexual flowers of *Chelyocarpus* are covered by thousands of Coleoptera. *Homalinotus nodipennis* (Cholinae) perforates the endosperm of the fruit of *Chelyocarpus repens* (see Fig. 6 in Kahn and Mejia 1988); *H. praelongus* is very abundant on the stem of *Astrocaryum* cf. *chonta* (Couturier 1988). Several species of *Homalinotus* were previously reported on palms by Lepesme (1947). Palm stems are frequently occupied by larvae of *Rhynchophorus palmarum* on *Mauritia flexuosa* and *Jessenia bataua* (Balick 1986), on *Aphandra natalia* (Borgtoft Pederson and Balslev 1990), *Metamasius hemipterus*, and *Rhinosthomus barbirostris* on *Mauritia flexuosa* (Urrego Giraldo 1987).

Several species of Chrysomelidae, such as *Demotispa* sp. on *Oenocarpus balickii*, were encountered on palm fruits; the following species were found on leaves of *Astrocaryum* cf. *chonta*: nymphs and adults of *Spaethiella* sp. (Cassidinae) on the pinnae, and larvae of *Cephaloleia* sp. and *Alurnus humeralis* (Hispinae) at the base of the pinnae in bud. Larvae of several species of Scarabaeidae, and adults of *Harposcelis paradoxa* live in the thick, spiny muff formed by the persistent sheaths of dead leaves of several species of *Astrocaryum* (all of the subgenus *Monogynanthus* section *Ayri*).

Cyclocephala amazona is frequent on staminate flowers of various palms. Larvae of *Macrodontia cervicornis* (Cerambycidae) have been observed in sheaths and petioles of *Maximiliana maripa* (G. Tavakilian, pers. comm.). Other Coleoptera were also found on native palms: two species of Elateridae, both with larvae and adults, *Halolepta* spp., and *Oxysternus maximus* (Histeridae) which is known to be a predator of *Rhynchophorus palmarum*.

Hemiptera. Bugs of the genus *Lincus* (Hemiptera Pentatomidae Discocephalinae) have been found on Amazonian native palms (Couturier and Kahn 1989 and unpublished data): *Lincus malevolus* on *Astrocaryum* cf. *chonta* in the lower Ucayali River valley and *A. scopatum* in the upper Marañon River valley, *Lincus hebes* on *A.* cf. *chonta* in Madre de Dios, *Lincus spurcus* on *A. carnosum* in the upper Huallaga valley, these three species in Peru; *Lincus* sp. on *A. gratum* in Beni, Bolivia. The presence of imago, larvae, and eggs on the palms show that the bugs carry out their whole biological cycle there. *Lincus* spp. were found on sexual parts as well as inside the sheaths of the intermediate and lowest green leaves of the crown, among the spines on the back of the petiole. The bug is brown-black, as are the spines, making the insect difficult to detect on sterile palms, except by its odor. The frequency and density of bugs on *Astrocaryum* palms is very high; *L. malevolus* is also encountered on *Elaeis oleifera*, mainly when this species is flowering and/or fruiting, in a forest where *Astrocaryum* cf. *chonta* is abundant, most of these with a high density of the bugs (Llosa et al. 1990). These Pentatomidae are considered vectors of *Phytomonas* palm diseases: "marchitez sorpresiva" of the African oil palm, *Elaeis guineensis* (Desmier de Chenon et al. 1983; Desmier de Chenon 1984; Perthuis et al. 1985), and hartrot of the coconut tree, *Cocos nucifera* (Desmier de Chenon et al. 1983, Louise et al. 1986).

Various species of Reduviidae, which are animal predators, were found in *Astrocaryum* cf. *chonta* and *A. carnosum*. An infrequent species of Coreidae, *Anasa siblica*, and some species of Lygaeidae and Pyrrhocoridae (*Dysdercus* sp.) were occasionally caught on palms (Couturier 1988).

Homoptera. Some species of scale insects and mealy bugs were found on palms, among them two species of *Limacoccus* (Phoenicococcidae) which develop between the pinnae in the leaf bud of *Astrocaryum* cf. *chonta* and of *Orbignya polysticha* (Couturier 1988). Henderson (1990) reported plant hoppers (Fulgoroidea) on *Iriartella setigera* on leaves, and adult and juvenile tree hoppers, *Membracis* sp. (Membracidae) on the branches of the inflorescence. According to this author, Homoptera are common visitors with ants to the infructescence of *Socratea*.

Orthoptera Acrididae. Descamps (1984) reported 18 species of Copiocerae and Eucopiocerae (Copiocerinae Copiocerini) related to palms. Four species of Copiocerae, the distribution area of which extends from Honduras to the Atlantic Coast of Brazil and northern part of Argentine, were found on *Euterpe oleracea*: *Copiocera portentosa*, *C. surinamensis*, *Copiocerina* sp., and *Eumecacris* sp. Most Eucopiocerae live on palms, such as *Chapulacris palmicola* and *Halffterina furculata*; this group is only known from central American forests (C. Amedegnato, in prep.). In the subfamily Romaleinae, the tribe Leguini includes several species which present morphological convergences with Copiocerini and which are also related to palms according to Amedegnato and Poulain (1986).

Lepidoptera. *Castnia daedalus*, the biology of which was studied by Korytkowski and Ruiz (1979) in Peruvian Amazonia, is frequently found in a plantation of African oil palms in the upper Huallaga River valley; many larvae of this species and other Castniidae were caught by Couturier (1988) on the native *Astrocaryum carnosum*, in the neighboring forests. On native palms, this author also collected caterpillars of *Herminodes* sp. (Noctuidae), frequent on *Astrocaryum* cf. *chonta* and *A. carnosum*, of the leaf feeders *Brassolis sophorae* (Brassolidae), of Limacodidae, and of other groups of Lepidoptera.

Others Insects. Ants and termites are frequently found between the leaf sheaths, and on the infructescences; and wasps, which usually install their nest beneath the blade, are common on large-leafed palms. Henderson (1990) observed that nests of ants are commonly found in *Socratea exorrhiza* under the stilt roots where the latter leave the stem, and reported the case of *Azteca* sp. (Formicidae) on that palm species in Ecuador. This author also found *Ectatoma tuberculatum*, *Hypoclinea* sp., *Odontomachus mayi*, and *Paratrechina* sp. (Formicidae) on *Iriartella setigera*, all apparently attracted by homopteran exudates.

Others Arthropods and Vertebrates. Various species of spiders, scorpions, centipedes, toads, frogs, lizards, snakes, and rats live or shelter on many palms, mainly in species with persistent bases of dead leaves on the stem under the crown, such as in *Astrocaryum*, *Leopoldinia*, *Maximiliana*, *Scheelea*, or in the fibrous sheaths of living leaves, as in *Jessenia bataua*, *Mauritia carana*, *Oenocarpus* spp. Gasc (1986) inventoried amphibians and reptiles in the subterranean-stemmed, large-leafed palm, *Astrocaryum paramaca*, in a French Guianan forest where this species is particularly dense (88–195 palms above 2 m in height per hectare). He found 52 amphibia of 9 species on 44 palms, 84 reptiles including 68 sauria of 9 species on 58 palms and 16 snakes of 9 species on 15 palms. Pires (pers. comm. in Henderson 1990) reported that parrots made a nesting hole in swollen section of the stem of *Iriartea deltoidea*. Bats used the large leaves of subterranean-stemmed palms as well as those of acaulescent juveniles of several arborescent species as a shelter, under which they rest and eat fruits;

Fig. 61. *Orbignya spectabilis*, a subterranean-stemmed, large-leafed palm. The unarmed petioles collect dead leaves and ramlets

the seeds of *S. exorrhiza* are known to accumulate on the soil under such palms (Charles-Dominique 1986; Timm 1987; Sist 1989a).

6.4 Palms as Litter Collectors

De Granville (1977) described *Astrocaryum paramaca* and *A. sciophilum* as collecting dead leaves fallen from forest trees. Such a process was reported by Kahn (1983) for *Astrocaryum sociale*, *Scheelea* spec. nov. and *Orbignya spectabilis* (Fig. 61), in central Amazonia, and *Bactris rhaphidacantha* in French Guiana (Fig. 62). It also occurs with *Astrocaryum macrocalyx*, *Chelyocarpus repens*, and *Orbignya polysticha* in the understory of Peruvian terra firme forests. All adults of understory species with large leaves (4–7 m) forming a funnel-like crown are liable to collect litter. This aptitude is due to

Fig. 62. *Bactris rhaphidacantha*, a short-stemmed palm with 3–4 m-long leaves. The spines on the petioles contribute to catch more dead leaves fallen from the canopy

the generally low growth rate of understory species, and also depends on the size and the morphology of the palm leaves. For instance, subterranean-stemmed, large-leafed species with short petioles (*Astrocaryum sociale*, *Orbignya polysticha*, *O. spectabilis*) tend to retain more dead leaves at the basal insertion of the pinnae on the rachis toward the sheath than *Astrocaryum acaule* or *Scheelea insignis*, the leaves of which are erect with a 1–2-m-long petiole. On the other hand, strongly armed petioles of *Astrocaryum* spp. or *Bactris rhaphidacantha* are more efficient than the unarmed petioles of *Orbignya* and *Scheelea* in retaining dead leaves.

The density of these species can be very high; 676 subterranean-stemmed palms (height > 1 m) were counted in a central Amazonian forest on 0.72 ha (Chap. 3, Table 18). This strongly suggests that these funnel-like collectors influence the pattern of litter distribution. Moreover, these accumulations of dead leaves are invaded by roots from neighboring trees and shrubs (Kahn 1983).

6.5 Discussion

Palms are important components in the functioning of forest ecosystems, mainly due to their abundance and their very numerous interactions with other organisms. Moreover, most arborescent palms are highly productive plants and constitute an important source of litter. The case of *Mauritia flexuosa* is particularly noteworthy: the soils of *Mauritia* formations are essentially composed of dead material fallen from the tall palms. Even though the production of understory species as well as of juvenile forms of large palms is low (one to two leaves per year), the volume of litter they produce is far from negligible because of their very high densities. For the same reason, palm fruits constitute an important food stock for animals; many are those which eat them, and contribute to their dispersion in most cases. Some life forms of palms offer habitats for many animals, mainly insects, spiders, and scorpions, but also small vertebrates, such as toads, snakes, and rats. This is particularly the case with medium-sized palms, the trunk of which is covered by a muff of the persistent sheaths of the dead leaves. Understory species with large leaves, such as both medium-sized and subterranean-stemmed species, are dead leaf collectors. When abundant, they are likely to condition both litter and root distribution pattern in a significant way.

Most data presented here are too punctual to be generalized. Systematic counting is necessary to develop conclusions more thoroughly. Comparative studies of the role played by palms in forest ecosystem functioning must be conducted with arborescent and understory palms, taking into account their different life forms. How different are the patterns of litter accumulation, rainfall interception by shrub foliage, and arthropod distribution in a forest with an understory dominated by small, multistemmed palms, and in a forest with an understory dominated by subterranean-stemmed, large-leafed palms? However significant or insignificant the difference may be, such an answer will be a very interesting contribution to our understanding of forest ecosystem dynamics.

7 Palms and Forest Management in Amazonia

7.1 Uses and Economic Potential of Amazonian Native Palms

A world without palms is a nonsense for most Amazonian inhabitants. Many data on ethnobotany and economic botany of palms have been published since Wallace's book (1853), *Palm Trees of the Amazon and Their Uses*: Lévi-Strauss 1952; Jordan 1970; Wheeler 1970; Dugand 1972; Cavalcante 1974; Schultes 1974, 1977; Johnson 1975, 1982; Kitzke and Johnson 1975; Anderson 1978, 1988; Balick 1979a, b, 1981, 1984, 1985a, b, 1986; Putz 1979; Mejia 1983, 1988; Balslev 1987; Kahn and Mejia 1987; Boom 1988; Balée 1988; Coradin and Lleras 1988; Lleras and Coradin 1988; Strudwick and Sobel 1988; Borgtoft Pederson and Balslev 1990; Kahn 1991; Barfod et al. 1990. Palms provide many useful products. In fact, all parts of the plants are used: leaves (thatching, basketry, building materials, wax, fibers), trunk (building material, starch), apical meristem (palm heart), fruit (edible fruit, oil, charcoal, vegetable ivory), and even roots (medicines). A few species have economic potential as edible fruit, palm heart for canning, oil, fiber, and starch, or constitute a gene bank for genetic improvement of cultivated or promising native species. They are mainly large, arborescent palms which can offer ample quantities of their products due to their size and high productivity.

7.1.1 Native Palms of Significant Economic Importance

Most of the following species provide products important for local trade:

Acrocomia lasiospatha is an arborescent, single-stemmed palm. Its distribution in Amazonia is restricted to the eastern coastal region. The mesocarp of the fruit is eaten. Lleras and Coradin (1983) consider this species as a promising source for combustible oil; the potential oil yield in natural stands is between 2.5 to 5 t/ha.

Aphandra natalia is a medium-sized, single-stemmed palm. Its distribution is limited to western Amazonia, in Ecuador, where it forms dense populations. The mesocarp is eaten when it is still fleshy and yellow, the fluid endosperm is drunk, and the palm heart is edible; fibers are extracted from the leaf sheaths and petioles are used for manufacturing brooms (Balslev and Barfod 1987). One individual may yield up to 4.5 kg of fiber in 1 year (Borgtoft Pederson and Balslev 1990). These authors considered it a promising species for agroforestry systems.

Astrocaryum aculeatum is an arborescent, single-stemmed palm, frequent in deforested areas in central Amazonia. The mesocarp is eaten and fruits are sold in the streets and markets of Manaus.

Astrocaryum chambira is a large, single-stemmed palm frequent in western Amazonia. The endosperm is drunk or eaten as with coconut. The epidermis of the pinnae of young leaves is used for fiber (Wheeler 1970; Schultes 1977) which serves to make hammocks, and bags called "shicras" in Peru (Mejia 1988).

Astrocaryum vulgare is an arborescent, multistemmed palm, 15–20 m in height. The mesocarp is rich in provitamin A. It provides a fat, mashed pulp which is used to prepare the very popular French Guianan "bouillon d'awarra", traditionally eaten at Easter. Fiber is occasionally extracted from the pinnae (Schultes 1977). Palikur Amerindians prepare a decoction of the roots which is said to have an effect against furunculosis, and use oil extracted from the seed to cure boils and toothache (Grenand et al. 1987).

Bactris gasipaes is a medium-sized, multistemmed species. It is cultivated throughout the Amazon basin and tropical South America, and has an incontestable economic importance providing edible fruit, oil, and palm hearts for the canning industry, as well as food for animals (Mora-Urpi 1983; Clement and Mora-Urpi 1987; Clement 1988). This species constitutes a major potential component of agroforestry systems. However, its origin is uncertain, and natural populations were not found in Amazonia. The reader should therefore not be surprised that we omit this important palm while dealing with the management of Amazonian forest ecosystems for native palms.

Desmoncus spp. are climbing, multistemmed palms. Exploitation as rattans is incipient as yet. A small factory has developed in Iquitos, and the feasibility of a rattan farm is being analyzed (K. Mejia, pers. comm. 1990).

Euterpe oleracea is an arborescent, multistemmed species. The fruits provide a popular drink and a flavoring for ice cream from the mesocarp. In the Brazilian state of Pará, the drink is locally called "vinho do assai". It is an important dietary component and, consequently, an important item in local trade; 54 507 tons of seeds were produced in 1979 (Coradin and Lleras 1983). This palm is also used in the palm heart canning industry; 93% of the national production is from Pará (Johnson 1982). Strudwick and Sobel (1988) produced a detailed and well-illustrated study of the uses of the species and pointed out its economic importance in eastern Amazonia. *E. oleracea* is one of the keys for tidal várzea management in the Amazon estuary (Anderson et al. 1985; Anderson 1988). Natural populations of this species have been more or less successfully exploited for heart canning for the last 10 years in French Guiana (lower Approuague and Oyapock River). The Guianan Agroindustrial Company (CAIG) has been exporting an average of 800 000 cans yearly for the period 1985–1986; this production corresponds to more than 2 million palm hearts (Ricci 1990). Root sap of *E. oleracea* is used as salt and palm heart pressed or dried is said to be a good cicatrisant (Grenand et al. 1987).

Euterpe precatoria is an arborescent, single-stemmed species. The fruits are used in the preparation of a "vinho do assai" in central Amazonia (Brazil). In Peru, however, where the palm is abundant, the fruits are not consumed, but the palm is cut down to extract the heart, even though it is a single-stemmed species. Most restaurants serve it as "ensalada de chonta", and a canning factory exploits the species. As a result, dense populations of *E. precatoria* are ever more distant from Iquitos. The trunk is split to make the walls of rural houses (López Parodi 1988). The root is crushed and boiled and the decoction is taken to treat malarial fever (Schultes and Raffauf 1990).

Jessenia bataua subsp. *bataua* is an arborescent, single-stemmed palm. An oil of very high quality, similar to olive oil, is obtained from the pulp of the fruit (Moldenke 1949; Pinto 1951; Pesce 1985; Balick 1986). The presscake provides highly nutritional protein (Balick and Gershoff 1981). An adult tree produces up to four infructescences per year and a single infructescence may contain 1200–2200 fruits. In Peruvian Amazonia, the annual production most often ranges from one to two bunches. In spite of a relatively low growth rate, this species is considered one of the most promising native palms. The wood is also used for cabinet work because of its ornamental aspect and its hardness. Schultes and Raffauf (1990) reported that all Amazonian natives use the oil from fruit as an antitubercular medicine and the Waoranis Indians in Ecuadorean Amazonia employ the adventicious roots in the treatment of worms, diarrhoea, headache, and stomach ailments.

Mauritia flexuosa is a large, single-stemmed palm, which can reach over 30 m in height. It is a dioecious species; staminate and pistillate flowers are on separate plants. It is particularly exploited for its fruit, which constitutes an important trade in Peruvian Amazonia (Padoch 1988). The orange pulp is eaten or is used to prepare a drink, cakes, and ice cream. Fruit productivity is high (see Sect. 6.2.2). Fruit yield was estimated to be 6.5 mt/ha in the lower Ucayali River (C. Peters, pers. commun. 1987) and at 9.07 mt/ha in Colombian Amazonia (Urrego Giraldo 1987). Lleras and Coradin (1988) consider *M. flexuosa* as very promising palms for oil production. This species also provides useful fiber (Schultes 1977; Balick 1984); the leaves are used for thatching, the petiole currently serves as building material and has been proposed as source for paper production (De los Heros and Zárate (1980/1981). Moreover, the trunk contains up to 60% of its dry weight as starch. As in many palms, when cut down, larvae of curculionids of the genus *Rhynchophorus* develop in the trunk (see Sect. 6.3.3). Locally named "suris" in Peru, they are a favorite food item. Up to 500 edible larvae of *R. palmarum* may live in a single decaying trunk (Borgtoft Pederson and Balslev 1990). The density of *Mauritia flexuosa* is very high. On one 1-ha plot surveyed, adult palms represented 60% of the trunked individuals; applying this rate to other data (see Sect. 3.3.4), a range of around 130 to 200 adult plants per ha can be estimated. A comparison with the 143 plants per ha in an African oil palm plantation underscores the very high densities reached by *M. flexuosa* formations. Moreover, these formations are extensive (Spruce 1871; Bouillenne 1930; Moore 1973; Klinge et al. 1990), especially in the Amazon and Orinoco estuaries and in the Peruvian region of the basin where they cover 21 and 34% of two areas located near Iquitos and Nauta, respectively (Anonymous 1977, from Landsat data analysis). *Mauritia flexuosa* is a particularly promising species in Peru (Salazar 1967).

Maximiliana maripa is an arborescent single-stemmed palm. According to Blaak (1983), the properly ripened fruit of *M. maripa* contains about three to five times (15.4%) more oil in the mesocarp than has been mentioned in the literature (3.34%). Up to 12% oil/infructescence could be obtained by using seed from the best forest palms. Fruits (mesocarp) are eaten in French Guiana and sold in local markets.

Orbignya phalerata is an arborescent, single-stemmed palm. The kernel provides oil which represents 3.5% of the fruit weight. Presscake from kernel and fruit offers 11% protein. Natural stands could produce up to 100 litres of oil per hectare (Anonymous 1983, p. 24). Babassu oil seed production in Brazil was 236755 tons in 1977 (Johnson 1982). Babassu provides "roof thatch and building materials; income through extraction and sale of the oil-rich kernels; food through domestic consumption of vegetable oil, animal ration, and the heart of the palm; and fuel in the form of charcoal made from the fruit husks" (Anderson 1983, p. 158).

Phytelephas macrocarpa is a moderate-sized, multistemmed palm. The leaves are commonly used for thatching, and also provide fiber. When the fruit is unripe, the liquid, or later, gelatinous, endosperm is eaten; when the fruit is ripe, the endosperm becomes very hard and is known as "vegetable ivory", from which buttons and other small objects are made. Fruits and seeds are sold in the regional markets. In Ecuador, the "Golden Age" of vegetable ivory trade was the period 1885–1935 with a depression during World War I; the demand started to decline with World War II (Barfod 1989). As described by Barfod et al. (1990), the vegetable ivory industry is thriving again in this country.

7.1.2 Native Palms Important to the Gene Bank

Elaeis oleifera is a single-stemmed palm with a creeping trunk. Populations of this species are sources of genes for further improvement and hybridization with the African oil palm, *Elaeis guineensis*. The American species offers an oil of better quality, a lower height because of its creeping trunk, and a better resistance to diseases (Meunier 1976; Ooi et al. 1981). This species, which was found some years ago in Ecuadorean and Peruvian Amazonia (Balslev and Henderson 1986; Kahn and Mejia 1986), occurs throughout the basin.

Oenocarpus bacaba, *O. balickii*, *O. distichus*, *O. minor* and *O. mapora* – the three former are single-stemmed, the two latter are multistemmed species; *O. bacaba* and *O. distichus* are large palms, *O. balickii* and *O. mapora* are medium-sized species, *O. minor* is an understory species. Four other species, *O. circumtextus*, *O. discolor*, *O. macrocalyx*, and *O. tarampabo*, are very uncommon. As part of the complex *Jessenia-Oenocarpus* (Balick 1986), these species must be considered as part of the gene bank available for further improvement of *J. bataua*. Palm heart pressed or dried is used as cicatrisant as with *E. oleracea* in French Guiana (Grenand et al. 1987).

Scheelea spp. are large, single-stemmed palms. The leaves are commonly used for thatching, and the fruits to make charcoal. Belonging to the complex *Attalea-Orbignya-Scheelea-Maximiliana*, they are considered to have importance for the future improvement of the babassu palm.

7.1.3 Useful Native Palms of No Current Economic Importance

Although not providing products of any foreseeable economic potential, the following species have a significant place in the daily life of most inhabitants of the Amazon basin.

Astrocaryum jauari is an arborescent, multistemmed species. The fruits are used as bait for fishing (Wallace 1853), and the leaflets for basketry and fibers. Some attempts are made in Brazil to exploit the species for palm heart canning; Barcelos is a center of production. The trunk is rot-resistant and serves as building material in Peru (Mejia 1988). Its distribution, which is limited to the periodically flooded river banks, as well as the many spines on the trunk, reduce its potential as an economic plant. However, Borgtoft Pederson and Balslev (1990) propose this species as a potential component of agroforestry systems.

Astrocaryum murumuru is a medium-sized, multistemmed palm, strongly armed with large spines. The fruit is edible and has been used for oil production in eastern Brazil (Coradin and Lleras 1983; Pesce 1985; Lleras and Coradin 1988).

Dictyocaryum ptariense is an arborescent, single-stemmed palm. The trunk is occasionally used to make floors, walls, and doors of rural houses.

Geonoma spp. and *Hyospathe elegans* are understory, clustered palms. Their leaves are generally used for thatching and for making hats locally named "catouris" in French Guiana.

Iriartella setigera, *I. stenocarpa* are small, multistemmed palms. They were traditionally used by Indians to make blowguns (Wallace 1853; Spruce 1871). This use is starting again with tourism activities in Amazonia.

Iriartea deltoidea is a tall, single-stemmed palm. The trunk is split and the outer part is used as building materials, floor, doors, and walls of rural houses. A floor made of this material is said to last up to 6 years. The trunk traditionally provided material to make blowguns, bows, harpoons, and arrow parts (Henderson 1990). These objects are now being made for the tourist market.

Leopoldinia spp. are medium-sized, single or multistemmed palms. Leaf sheaths are highly fibrous. The fiber of *L. piassaba* is used for making ropes and brooms, and is currently exported from the upper Rio Negro valley (Putz 1979). Its leaves are commonly used for thatching.

Lepidocaryum tessmannii is an understory, small, multistemmed palm. The leaves are commonly used for thatching in Peruvian Amazonia (Kahn and Mejia 1987).

Manicaria spp. are moderate-sized, single or multistemmed palms. Their huge, 8-m-long, entire leaves provide robust thatching material. Wilbert (1980) reported the Warao Indians in Venezuela as using *M. saccifera* leaves as sails. The inhabitants of the lower Oyapock French Guianan region know that the liquid endosperm of unripe *M. saccifera* fruit has a diuretic effect (Grenand et al. 1987).

Mauritiella aculeata and *M. peruviana* are medium-sized, multistemmed palms. The pulp of the fruit is eaten as is that of *Mauritia flexuosa*.

Socratea exorrhiza is a tall, single-stemmed palm. The trunk is split for making house walls, doors, and floors. Pieces of stilt roots are sold in the regional markets of medicinal products in Brazil. The sap is known to be corrosive (Grenand et al. 1987).

Wettinia augusta and *W. maynensis* are both medium-sized palms frequent in western Amazonia. The stem is split and the outer part is divided into 3-m-long pieces which serve to attach leaves of small palms by the petiole and make cover units for thatching; such cover units are sold in regional markets. The outer part of the stem is also used to make arrow points, harpoons and bows which are sold in the tourist market.

Many other species are occasionally used, such as most species of *Bactris* spp., *Catoblastus drudei*, *Syagrus inajai*; the roots of the latter are said to cure measles (Grenand et al. 1987). Only a few genera are not known to have any use (Table 64).

Table 64. Amazonian palms and their most common uses[a]

	Species	Uses
1.	*Acrocomia lasiospatha*	Edible fruit[b]
2.	*Aiphanes* spp.	Minor use
3.	*Ammandra dasyneura*	?
4.	*Aphandra natalia*	Vegetable ivory[b], edible seed[b], fiber[b]
5.	*Asterogyne guianensis*	?
6.	*Astrocaryum aculeatum*	Edible fruit[b], building materials
	Astrocaryum chambira	Edible fruit[b], fiber[b]
	Astrocaryum jauari	Basketry[b], fiber
	Astrocaryum murumuru	Edible fruit, oil
	Astrocaryum vulgare	Edible fruit[b]
7.	*Attalea tessmannii*	Roof thatch
8.	*Bactris gasipaes*	Edible fruit[b], palm heart[b], oil[b]
	Bactris spp.	Minor use
9.	*Barcella odora*	Roof thatch
10.	*Catoblastus drudei*	Minor use
11.	*Chamaedorea* spp.	Minor use
12.	*Chelyocarpus* spp.	Roof thatch
13.	*Desmoncus* spp.	Rattans
14.	*Dictyocaryum ptariense*	Building materials
15.	*Elaeis oleifera*	Oil
16.	*Euterpe oleracea*	Edible fruit[b], palm heart[b], building materials[b], medicine
	Euterpe precatoria	Palm heart[b], building materials[b], edible fruit
17.	*Geonoma* spp.	Roof thatch[b]
18.	*Hyospathe elegans*	Roof thatch
19.	*Iriartea deltoidea*	Building materials[b]
20.	*Iriartella* spp.	Minor use
21.	*Itaya amicorum*	Roof thatch
22.	*Jessenia bataua*	Edible fruit[b], oil[b], cabinet work[b], medicine
23.	*Leopoldinia* spp.	Fiber[b]
24.	*Lepidocaryum* spp.	Roof thatch[b]
25.	*Manicaria* spp.	Roof thatch[b], minor uses[b], building materials[b]
26.	*Mauritia carana*	Fiber[b]
	Mauritia flexuosa	Edible fruit[b], fiber[b], roof thatch
27.	*Mauritiella aculeata*	Edible fruit[b]
28.	*Maximiliana maripa*	Oil, roof thatch[b], building materials
29.	*Oenocarpus* spp.	Edible fruit[b]
30.	*Orbignya phalerata*	Oil[b], charcoal[b], building materials[b], fiber[b], animal ration[b]
	Orbignya polysticha	Roof thatch[b]
31.	*Pholidostachys synanthera*	Roof thatch
32.	*Phytelephas macrocarpa*	Seed as edible and vegetable Ivory[b], fiber[b]
33.	*Prestoea*	Palm heart
34.	*Raphia taedigera*	Thatch
35.	*Scheelea* spp.	Oil, charcoal[b], building materials[b], animal ration
36.	*Socratea exorrhiza*	Building materials[b], medicine
	Socratea salazarii	Minor use
37.	*Syagrus inajai*	Minor use
	Syagrus sancona	Building materials
38.	*Wendlandiella* spp.	?
39.	*Wettinia augusta*	Minor use[b]

[a] This list, which is far from being exhaustive, points out that only 3 of the 39 native genera have no reported uses. One is prostrate, the two others are erect palms; all three have a very restricted distribution.
[b] Very common use. "Minor use" includes products for tourist trade (bows, arrows, blowguns, small objects, etc.).

7.2 Useful Palms in Forest Ecosystems

7.2.1 Terra Firme Forests

As concluded in Chapter 3, terra firme forests are characterized by the very low density of arborescent adult palms. Most of the useful species are arborescent palms. Therefore at first glance, this ecosystem does not seem to offer great potential from its palms (Table 65A). In the southeastern part of the Amazon basin, however, the babassu, *Orbignya phalerata*, is particularly abundant. This species has many uses and is considered to be one of the most promising species for eastern Amazonia (Anderson 1983; Anonymous 1983; Balée 1988; Anderson et al. 1991). It grows in terra firme forests, and its high density may be the result of long use of its populations by humans. *Orbignya phalerata* and most terra firme arborescent palms, such as *Astrocaryum aculeatum*, *A. chambira*, *A. vulgare*, *Maximiliana maripa*, *Oenocarpus bacaba*, are frequent and form dense populations in secondary vegetation (Table 65B). These species can be integrated into agroforestry systems (May et al. 1985b; Kahn 1988b; Borgtoft Pederson and Balslev 1990). *Geonoma*, *Hyospathe*, and *Lepidocaryum*, which are the smallest useful species, reach very high densities in terra firme forests; they represent only an occasional money input from thatching and the production of articles for the tourist trade.

7.2.2 Forests on White-Sandy Soils

On dry, orthic podzol, palm diversity and density is too low for these plants to have a significant economic potential (Table 65C). On gleyic podzols, which are waterlogged after rainfall, one species, *Jessenia bataua* subsp. *bataua*, is largely dominant with densities of more than 100 adult palms/ha. These natural stands would allow the management of such poorly drained, white-sandy soils (Table 65D).

7.2.3 Flooded Forest

Palms are present in all flooded forest ecosystems. The species composition, species richness, density – and as a result the economic potential – vary, however, from one ecosystem to another (Table 66). Only *Euterpe oleracea*, *E. precatoria*, *Oenocarpus mapora* are frequently found in forests on alluvial soils periodically flooded by whitewater, but they rarely form dense populations. Moreover, these soils are particularly suitable for crops, and the

Table 65. Useful and economic palms in unflooded forests of Amazonia

	A		B		C		D	
	f	d	f	d	f	d	f	d
Palms with significant economic potential								
Aphandra natalia (W)	?		L	H	–	–	–	–
Astrocaryum chambira (W)	L	L	H	H	–	–	–	–
Astrocaryum aculeatum (C, E)	L	L	H	H	–	–	–	–
Astrocaryum vulgare (E)	–	–	H	H	L	L	–	–
Bactris gasipaes[a] (T)	–	–	H	H	–	–	–	–
Jessenia bataua subsp. *bataua* (T)	M	L[b]	–	–	–	–	H	H[c]
Jessenia bataua subsp. *oligocarpa* (E)	L	L	–	–	–	–	–	–
Orbignya phalerata (E)	H	H	H	H	–	–	–	–
Palms important to the gene bank								
Oenocarpus bacaba (E, C)	L	L	M–H	M	–	–	–	–
Oenocarpus distichus (E)	L	L	M–H	M	–	–	–	–
Oenocarpus minor (C)	H	M	–	–	–	–	–	–
Useful palms without economic importance								
Bactris spp. (T)	H	H	–	–	L	L	M	M
Geonoma spp. (T)	H	H	–	–	–	–	M	M
Hyospathe elegans (T)	M	L–H	–	–	–	–	M	M
Iriartea deltoidea (W)[d]	H	M–H	–	–	–	–	–	–
Iriartella setigera (C, W)	M	L	–	–	–	–	M	M
Lepidocaryum spp. (C, W)	H	H	–	–	L	L	H	H
Mauritia carana (C, W)	–	–	–	–	–	–	L	H
Pholidostachys synanthera (W)	H	H	–	–	–	–	H	H
Scheelea spp. (W)	L	L	H	H	–	–	–	–
Socratea exorrhiza (T)	H	M	L	L	–	–	H	M
Socratea salazarii (W)	L	L	–	–	–	–	L	L
Wettinia augusta (W)	M	L	–	–	–	–	–	–

A: terra firme forests on clayey, well-drained soils (acrisol, luvisol).
B: secondary vegetation on well-drained soils.
C: low vegetation on dry, orthic podzol.
D: forests on irregularly waterlogged, gleyic podzol.
[a] Cultivated species.
[b] In Ecuadorean Amazonia.
[c] In Brazil and Peru.
[d] On mountain slopes.
(f: frequency; d: density – H: high, M: medium, L: low; E: East; C: Central, W: West, T: throughout the basin).

Table 66. Useful and economically important palms in flooded forests of Amazonia. (After Kahn 1991)

	Periodically flooded forests				Permanently flooded forests		Seasonal swamp forests (upland valleys)	
	White water		Black water					
	f	d	f	d	f	d	f	d
Palms with significant economic potential								
Euterpe oleracea (E)	M	M	–	–	H	H	H	H
Jessenia bataua subsp. *oligocarpa* (E)	–	–	–	–	–	–	H	H
Euterpe precatoria (W, C)	M	M	–	–	H	L–M	H	H
Jessenia bataua subsp. *bataua* (W, C)	–	–	–	–	–	–	H	H
Mauritia flexuosa (T)	–	–	–	–	H	H	H	M
Phytelephas macrocarpa (W)	H	H	–	–	–	–	L	L
Palms important to the gene bank								
Elaeis oleifera (T)	–	–	–	–	L	H	L	H
Oenocarpus mapora (W, C)	M	M	–	–	H	M–H	H	M–H
Useful palms without economic importance								
Astrocaryum jauari (T)	M–H	M–H	M–H	M–H	–	–	–	–
A. murumuru (E)	H	H	–	–	–	–	L	H
Geonoma spp. (T)	H	H	–	–	H	H	H	H
Hyospathe elegans (T)	L–M	M–H	–	–	–	–	M	H
Iriartea deltoidea (W)	–	–	–	–	–	–	L	M–H
Leopoldinia spp. (C)	L	M	H	H	–	–	–	–
Manicaria spp. (C, E)	–	–	–	–	–	–	L	H
Mauritiella spp. (T)	–	–	–	–	–	–	L	M–H
Scheelea spp. (W)	H	H	–	–	–	–	–	–
Socratea exorrhiza (T)	M	M	–	–	H	M–H	H	M–H

f: frequency; d: density – H: high, M: medium, L: low; E: East, C: Central, W: West, T: throughout the basin.

native palms, though highly productive plants, will not be able to compete economically with most crops, without genetic improvement. However, there are dense stands of *Euterpe oleracea* in the Amazon River estuary on tide-driven floodplains which have been successfully managed (Anderson et al. 1985; Anderson 1988). There are no palms of significant economic importance in the forests which are periodically flooded by blackwater rivers; only *Astrocaryum jauari* and *Leopoldinia piassaba* may have economic potential, the former for palm heart canning (Borgtoft Pederson and Balslev 1990), the latter as fiber (Putz 1979).

Economically important palms, *Euterpe oleracea*, *E. precatoria*, *Jessenia bataua* susp. *bataua*, and *Mauritia flexuosa* form dense and extensive populations in the seasonal swamp forests of upland valleys, and on areas which are permanently flooded by standing water.

7.3 Palms as a Key in Swamp Forest Management

The exploitation of Amazonian forests generally leads to their destruction. Vast areas are transformed into agro-industrial or timber plantations, and numerous small areas, deforested by shifting cultivation or timber extraction, spread irreversibly with increasing human population density. Except those forests on periodically flooded alluvial soils which are traditionally cultivated, most flooded forests have not been severely affected by deforestation because extreme ecological constraints make them unsuitable for most crops and pastures. This is particularly the case of the swamp forests which are neglected in most plans for forest management.

Flooded forests cover around 2% of Brazilian Amazonia (Adis 1984). They represent 12% of Peruvian Amazonia (Salo et al. 1986); swamp forests cover as much as 36.4% of the Iquitos-Nauta-Requena region including the Marañon and Ucayali River confluences (Anonynous 1975). The Marañon subsidence, which extends to the foot of the Andes, is almost covered in swampy areas. On the other hand, there are no estimates of the total area covered by seasonal swamp forests in the upland valleys throughout the basin, but this is surely not negligible.

In the swamp forests, native palms form either (1) almost monospecific stands, e.g., *Euterpe oleracea* swamps in eastern Brazil and in the Guianan coastal plain and *Mauritia flexuosa* swamps throughout the basin; or (2) multispecific communities, e.g., the seasonal swamp forests of central and western Amazonia which are dominated by *Jessenia bataua, Mauritia flexuosa* and *Euterpe precatoria*. Such natural stands are the most common "oligarchist" forests defined by Peters et al. (1989). They differ from African oil palm or coconut tree plantations by the occurrence of natural regeneration (nil in a plantation) and by the within-species heterogeneity, which is high in natural populations and intentionally reduced in plantations. As a result, the management of such palm stands will present more similarities with agroforestry systems than with palm plantations. As pointed out by Altman and Cordeiro (1964) and by Anderson (1988) for *Mauritia flexuosa* and *Euterpe oleracea* stands respectively, the management of natural palm populations to be a viable land-use alternative requires a high degree of integration in the exploitation of all useful parts of the palms as well as of other accompanying trees. Extraction of products must be controlled in order to maintain the economic potential of the ecosystem. More-

Table 67. Sex ratio of *Mauritia flexuosa* populations in relation to collecting activities

	Males No. and %	Females No. and %	Sex unknown
With intensive collecting (Peru)			
[a] 1 ha surveyed	15 (83.3)	3 (16.7)	0
[a] 1 ha surveyed	100 (72.5)	23 (16.7)	15 (10.9)
[a] 0.5 ha surveyed	13 (68.4)	4 (21.1)	2 (10.5)
With former collecting activities (Peru)			
[b] Mean from 20, 0.5-ha plots	111 (61.7)	69 (38.3)	
Without collecting (Colombia)			
[c] Mean from 5 sites each with 2–3 samples of 5, 0.5-ha plots	140 (51.5)	132 (48.5)	

[a] Kahn (1988b).
[b] Salazar and Roessl (1977).
[c] Urrego Giraldo (1987).

over, product yield will be improved by (1) increasing palm density by controlling natural regeneration and introducing seedlings of promising species, and (2) selecting plants with higher productivity and higher product quality.

The management of natural palm stands will depend on certain changes at the socioeconomical level in order to reorganize efficiently the collecting activities, product transformation, and trade. Indeed, some customs will have to be abandoned and new practices introduced, as discussed in the following two examples:

Palm fruit collecting usually involves cutting down the tree, and this occurs throughout the Amazon basin. As a result, only 1 year's yield is generally harvested per palm, and the native vegetation is progressively losing its economic potential. In the case of the dioecious *Mauritia flexuosa* in Peruvian Amazonia, there is currently a proponderance of male plants among adult palms, reflecting past collecting practice (Table 67). In contrast, Urrego Giraldo (1987) found only a slight difference between male and female density (51.5 and 48.5% respectively) in a *Mauritia* formation in Colombian Amazonia, which has not been exploited. A similar problem exists for the single-stemmed *Jessenia bataua* and *Euterpe precatoria* which are becoming rare in most inhabited areas, while the multistemmed *Euterpe oleracea* is not endangered by cutting down the fruiting stems (Oldeman, 1969). De Granville (1985) has campaigned against this destructive fruit-collecting practice in French Guiana. Climbing methods have been described by Corner (1966) and Hartley (1977). Several climbing techniques are developed, such as the "palm tree climber" (Davis 1984); they could be introduced into Amazonia, and would make it possible to exploit a palm's lifetime yield. However, although everyone can cut down a palm with an axe, only an expert can climb daily to collect fruits. The introduction of these techniques implies a social reorganization of collecting activities: people will have to be trained with adequate material and fruit collecting must become a fulltime job. Currently palm extractivism is essentially an individual activity. On the regional scale, only a structuring into small enterprises or cooperatives will make it possible to control regeneration, improvement, extraction, product transportation, and marketing, i.e., an efficient and rational management of vast palm stands. Palms are productive; their regeneration is natural: it is man who must learn to organize himself to benefit from nature while avoiding its destruction.

All natural resources are not currently being exploited. The case of the starch contained in the trunk of *Mauritia flexuosa* is particularly obvious. In Amazonia, it was extracted only by the Warao Indians of the Orinoco delta (Heinen and Ruddle 1974). However, the exploitation of palm starch is an activity of some economic importance in southeastern Asia (Ruddle et al. 1978). The technologies for extracting starch must be introduced into Amazonia, mainly in the Peruvian part of the basin, in order to contribute to managing the vast areas covered by *M. flexuosa* swamps. Starch extraction will permit the use of male plants. However, this palm-destroying exploitation will have to be controlled in order to maintain a density sufficient for pollination of female trees and subsequent fruit production. A few male trees per hectare will probably ensure a high fertility rate.

The deforestation in Amazonia mainly affects the terra firme forests, that is, the forests which present the highest species diversity. Seasonal and permanent swamp forests have much lower diversities, and they are often dominated by only one or a few species. The management of native palms in these swamp forests could provide several products, such as oil, starch, food and drink, animal food, fiber, and cellulose, which, at present, are obtained from deforested areas of the uplands. In this way, it could contribute to limiting the destruction of species-rich terra firme forest ecosystems.

8 Conclusion

The Amazonian palm flora is highly diversified in species and life forms. Palm communities allow forest ecosystems to be characterized by their floristic composition, diversity, density, and structure; these are thoroughly different when located on well-drained soils or on waterlogged and/or flooded soils.

Richness of Palm Flora and Geographic Distribution. Of all palm genera of the world, 39 (ca. 20%) are native to the Amazon basin. A few genera are endemic to the Amazon basin: *Aphandra*, *Barcella*, *Iriartella*, *Itaya*, *Leopoldinia*, *Lepidocaryum*, *Manicaria*, and *Wendlandiella*. *Astrocaryum*, *Bactris*, and *Geonoma* are particularly diversified in Amazonian forests with about 20, 50, and 30 species, respectively. Most genera are represented by only one or a few species, however. The palm flora in the western region is particularly rich with 34 genera, of which 13 are endemic to this region. Two genera have been found only in the central region, and two others are limited to the eastern region.

Life Forms of Amazonian Palms. From only two growth models, Corner's and Tomlinson's, the Amazonian palms have developed a wide range of life forms. They are subacaulescent and small-leafed, or subterranean-stemmed and large-leafed, or they produce an aerial stem which is unbranched or branched basally, erect, prostrate, or climbing. Aerial-stemmed forms are arborescent with adult height over 12 m, or medium-sized, with adults from 6 to 12 m, in both cases with large leaves (>4 m) and big diameter (>20 and >12 cm, respectively), or shorter leaves and smaller diameter, or they are small with adult height less than 6 m. The largest palms reach more than 30 m in height, the smallest remain under 1 m.

Palm Communities in Forest Ecosystems of Amazonia. Diversity of species and genera reaches very high levels in terra firme forests. The number of species per area unit is higher in terra firme forests of the central and western regions, and the number of genera per area unit is clearly higher in the western forests, as shown by both following striking examples from Peru: 29 species in 16 genera in 0.71 ha; 34 species in 21 genera in 0.5 ha. Most species are small palms which never grow more than 10 m in height.

All medium-sized and small life forms – subacaulescent; subterranean-stemmed and large-leafed; climbing; erect or prostrate, single- or muti-stemmed – occur in the understory of terra firme forests. These and juveniles of arborescent species represent about 99% of the community. The density of arborescent palms above 10 m in height is very low. Arborescent, multistemmed palms are absent from the terra firme forests. The structure of the palm community is a function of forest architecture and dynamics. The development of large palms which requires a high level of light intensity depends on frequency and size of chablis. The density of most understory species and the distribution of tall palms are affected by local topography which: (1) modifies lateral drainage and water supply, (2) influences forest architecture and dynamics, and thus, light intensity in the understory, by conditioning the frequency and size of chablis.

Palm flora of flooded forests is more homogeneous throughout the basin and clearly less diversified than in terra firme forests. Nevertheless, these ecosystems are characterized by associations of palm species: the most characteristic are: *Euterpe precatoria*, *Jessenia bataua*, *Mauritia flexuosa* in the seasonal swamp forests of central and western regions; *Astrocaryum carnosum*/*A.* cf. *chonta*/*A. urostachys*, *Phytelephas macrocarpa*, and *Scheelea* spp. in forests on periodically flooded alluvial soils in the western region; and almost pure stands of *M. flexuosa*, throughout the basin, or *Euterpe oleracea*, in the eastern region, both species in swamp forests on permanently flooded organic soils.

The density of arborescent palms with a height over 10 m is very high in the swamp forests. These plants are the major components of the forest canopy. However, the density of tall palms is low in the forests on periodically flooded alluvial soils, where juveniles and adults of medium-sized species are concentrated between 1 and 10 m in height. In all flooded forests, density of seedlings and juveniles is high in spite of an annual 3- to 9-month flooding.

The study of the relationships between soils and palm distribution demonstrates the importance of water drainage and flooding. In inundated forests, the severe ecological conditions imposed by flooding maintain a highly specialized flora and a lower diversity as a result (Richards 1969). The species found on poorly drained and waterlogged soils develop roots which are specialized in absorbing oxygen in anaerobic conditions. Juveniles of most arborescent species which form dense populations on swampy soils occur in the understory of terra firme forests, but they rarely produce a trunk there. The canopy of the forests on swampy soils is more open than that of terra firme forests and the intensity of the light which reaches the understory is higher in the former than in the latter. These species do develop trunks, however, on well-drained soils in deforested areas and in secondary vegetation. Most species growing in terra firme forest understory are never found on waterlogged or flooded soils.

One of the major features which has come to light through the comparison of palm inventories carried out in the most characteristic forest ecosystems of the Amazon basin is the very high diversity and density of small species in the understory of terra firme forests, while juveniles of arborescent species are better represented in the understory of swamp forests. The most diversified palm genera, *Bactris* and *Geonoma*, both essentially composed of small species, are found in all forest ecosystems, but they present a peak of diversity in terra firme forest understory. This suggests that an important trend of adaptive radiation of palms throughout Amazonia has been to colonize the lower reaches of terra firme forests.

Why are terra firme forests so rich in palms and others plants? Ecological stability and low competition was emphasized by Van Steenis (1969), while Ashton (1969) explained the complexity of Far Eastern rain forests in terms of "seasonal and geological stability" and "their great age". The understory of terra firme forests is characterized by a stable environment of high humidity and low light intensity. It is certainly disturbed by tree falls, but the original microclimate is soon re-established by the development of pioneer trees. Such disturbance temporarily affects the size of understory species communities as well as the growth and reproductive rate of most species (Piñero and Sarukhán 1982; Chazdon 1984, 1986a). Understory species composition and the high diversity of the ecosystem are maintained on the whole.

Palms as Major Components in Forest Ecosystem Functioning. Most arborescent palms are highly productive plants. In swamp forests, palms are dominant and play a major role as source of organic matter. The production of leaf and sexual material of a *Mauritia flexuosa* stand is twice that of a terra firme forest, which may explain the maintenance of acidic, organic soils in these formations.

Palms also play an important role as habitat for many animals, including invertebrates and vertebrates. This role increases in importance with the increasing degree of persistence of the sheaths of dead leaves on the trunk and the increasing fibrousness of the sheath. Species which clearly influence the distribution of fauna in the forests includes: *Astrocaryum carnosum*, *A.* cf. *chonta*, *A. murumuru*, *A. sciophilum*, *A. urostachys* and *Phytelephas macrocarpa*; all these form very dense populations up to 10 m in height. Several insects (bugs, caterpillars) which are abundant in the sheaths of these palms are vectors of diseases or are predators of the African oil palm and the coconut tree. Many insects related to palms contribute as pollinators, and most vertebrates as seed dispersers.

Another role played by subterranean-stemmed, large-leafed palms is to collect dead leaves between the petioles. Because of the high density reached by this life form in some forests, the accumulation of dead leaves and woody material in palms is by no means negligible, and is likely to affect the pattern of litter distribution.

Palms and Forest Management. With the exception of the forests located in the southeastern region of the Amazon basin, which are dominated by the babassu palm, most Amazonian terra firme forests are characterized by low density of economically promising species. In forests on periodically flooded alluvial soils, the economic potential of native palms is reduced because of the high fertility of these soils and consequent intensive use for agriculture. Nevertheless, several palm species of terra firme forests and forests on alluvial soils offer products which are sold in the regional markets. These products include fiber, leaves for thatching, vegetable ivory, and some edible fruits, which provide a complementary income.

Swamp forests dominated by large palms offer the greatest economic potential. These natural populations of native palms must be taken into account in any plan for swamp forest management. This is particularly true for Peruvian Amazonia, where swampy areas are very extensive.

Palms and Conservation. The list of endangered New World palms provided by Johnson (1988) includes the following Amazonian species: *Attalea tessmannii*, *Oenocarpus circumtextus*, and *Itaya amicorum*. The two former, which are known only from a few specimens, may be extinct. The latter occurs in restricted areas. This is also the case of *Aiphanes ulei, Asterogyne guianensis* and *Chelyocarpus repens*. These species are endangered and will probably become extinct if their areas are deforested. Some species are not frequent, such as *Astrocaryum javarense*, *A. ulei*, *Dictyocaryum ptariense*, *Oenocarpus tarampabo*, *Syagrus stratincola* and are threatened by extensive deforestation. Nevertheless "there is some justification for optimism . . . we cannot confirm any total extinctions" (Johnson 1988, p. 271).

Supplement

In order to help the reader to identify Amazonian palm genera in the field, this supplement includes two keys based on vegetative characters, brief descriptions of Amazonian palm genera, and a list of the most common vernacular names of palms which are used in Amazonia (Kahn 1990a,b, slightly modified here).

1. Identification of Amazonian Palm Genera from Vegetative Characters

The first key, which deals with all palms with accessible leaves, i.e., palms less than 10 m in height, can be successfully applied to identify seedlings, juveniles and sterile palms at the genus level. The key starts with the morphology of the blade (Fig. 63): (1) palmate, "fan-like" (*Chelyocarpus*, *Copernicia*, *Itaya*, *Lepidocaryum*, *Mauritiella*, *Trithrinax*), or costapalmate, i.e., with a short, curved rachis in the blade (*Mauritia*), (2) blade entire and bifid, or only two segments or pinnae (seedlings of many genera and some adults of *Bactris*, *Chamaedorea*, *Geonoma*, *Wendlandiella*), (3) blade entire, not bifid (seedlings of several genera and adult form of *Manicaria*), and (4) pinnate, "feather-like" (seedlings, juveniles, and adults of most genera).

The forms of entire or bifid blade and of leaflets (Fig. 64) are dealt with next in the key. For instance, the pinnae, which are pointed at the tip or truncate and broad apically (wedge-shaped), allow the separation of *Aiphanes* and the Iriarteeae (*Catoblastus*, *Dictyocaryum*, *Iriartea*, *Iriartella*, *Socratea*, and *Wettinia*) from other genera. The presence or absence of spines is also used, together with: (1) the color of the underside (abaxial) of the blade (white in *Astrocaryum* and *Jessenia*, glaucous in *Acrocomia* and *Oenocarpus*, green in most genera, or green with brownish longitudinal stripes in *Attalea*, *Maximiliana*, *Orbignya* and *Scheelea*); (2) the form of the pinnae (linear, lanceolate, or S-shaped); (3) the tip of the pinnae, either symmetric (acute or slightly bifid) or asymmetric (oblique notched); and (4) the ribs (main nerves) prominent above and/or beneath. Other characters, such as the tubular or split sheath, the arrangement of the pinnae either in one plane or oriented in several directions, are considered next. In several

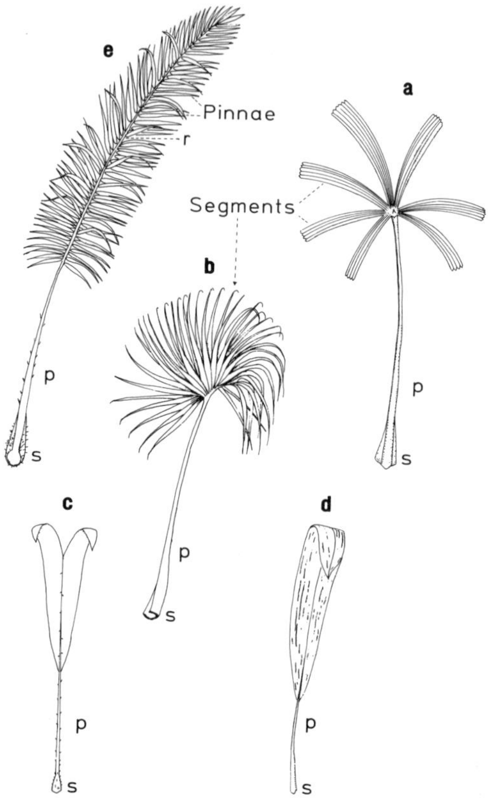

Fig. 63a–e. Leaves of Amazonian palms. **a** Palmate leaf. **b** Costapalmate leaf. **c** Leaf with entire, bifid blade. **d** Leaf with entire, not bifid blade. **e** Pinnate leaf. *p* petiole; *r* rachis; *s* sheath

cases, complementary characters are given to make the choice among dichotomic branches easier.

The second key deals with tall palms, the leaves of which are inaccessible. Characters of the leaves, of the trunk, of the roots, and of the crown physiognomy are used.

These keys are for use in the field, in all ecosystems of the Amazon basin, in primary and secondary forests. Forty-three genera are treated. Characters used to separate them are those of Amazonian native species, except for *Cocos* (*C. nucifera*), *Elaeis* (a native species, *E. oleifera*, and the introduced African oil palm, *E. guineensis*), and *Roystonea* (considered only in the second key). *Copernicia* and *Trithrinax* are included in the key in spite of not being truly Amazonian; they occur in the region adjacent to the basin and mark the southern limit of the Amazonian palm flora.

It was impossible to separate the seedlings and juveniles of the complexes *Attalea-Maximiliana-Orbignya-Scheelea*, *Chelyocarpus-Itaya* and

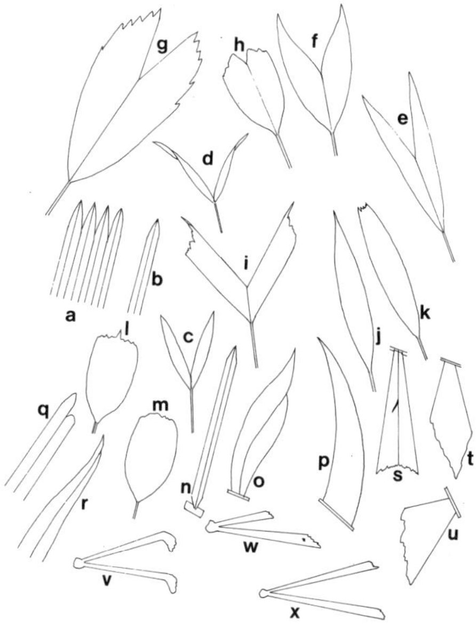

Fig. 64a–x. Leaflets of Amazonian palms. **a** Segment multi-pointed, several folded; **b** segment or pinna one-pointed; **c** two pinnae in a closed-V shape; **d** two pinnae in an open-V shape, each twisted outwards; **e** blade entire and bifid, each half with straight margins; **f** blade entire and bifid, each half S-shaped; **g** pinna or half of bifid blade regularly multi-pointed toward the tip; **h** blade bifid and divided for less than half its length, each half truncate or irregularly toothed; **i** blade bifid and divided for more than half its length, each half truncate or irregularly toothed; **j** blade entire, not bifid, far longer than wide, with tip pointed; **k** blade entire not bifid, far longer than wide, with tip truncate or denticulate, not sharply pointed; **l** blade entire, not bifid, margins rounded, with tip pointed; **m** blade entire, not bifid, margins rounded, with tip not pointed; **n** pinna straight, linear-lanceolate; **o**, **p** pinna S-shaped; **q**, **r** tip of pinna asymmetric, obliquely notched; **s** pinna truncate, with broad tip, usually with spines; **t**, **u** pinna wedge-shaped or like a fish fin, without spines; **v** upper parts of longitudinally divided pinnae slightly erect, tip drooping; **w** upper parts of longitudinally divided pinnae shorter than the lower parts, tip not drooping; **x** upper parts of longitudinally divided pinnae not shorter than the lower parts, tip not drooping (**v**, **w**, **x** leaf viewed in cross-section)

Phytelephas-Ammandra-Aphandra. Furthermore, some adult forms of the former could not be distinguished clearly at the genus level from vegetative characters. This was also the case with the related *Asterogyne-Geonoma-Pholidostachys*. The scope of the key will not be greatly reduced, because *Asterogyne* and *Pholidostachys* are represented by only one species, the former in a small area in French Guiana, the latter restricted to the western region as the species of *Ammandra* and *Aphandra* are.

Key 1: Leaves Are Accessible (Palms Less than 10 m in Height)

1. a Leaves palmate (fan-shaped) or costapalmate (with a short, curved rachis in the blade); segments three or more (Fig. 63a,b). [see 2]
1. b Leaves with only two segments or pinnae, or blade entire and bifid (Fig. 63c). [see 13]
1. c Leaves with blade entire; pointed, not bifid (Fig. 63d). [see 35]
1. d Leaves pinnate (Fig. 63e). [see 43]
2. a Leaves costapalmate; sheath diameter more than 30 cm. [*Mauritia*]
2. b Leaves palmate; sheath diameter less than 30 cm. [see 3]
3. a Blade divided into segments multi-pointed, several-folded (Fig. 64a) with several ribs, white beneath. [see 4]
3. b Blade divided into segments one-pointed (Fig. 64b), connate or not basally, each with one or several ribs, white, green or glaucous beneath. [see 5]
4. a Petiole split basally in the sheath; this splitting opposite the petiole (see old leaves). [*Itaya*]
4. b Petiole not split basally in the leaf sheath; this closed. [*Chelyocarpus*]
5. a With spines on the upper surface of the rib(s) (select the youngest leaves). [see 6]
5. b Without spines on the upper surface of the rib(s). [see 8]
6. a Blade white beneath; spines on the stem. [*Mauritiella*]
6. b Blade green beneath; without spines on the stem. [see 7]
7. a Segments one-ribbed, connate at base forming two groups separated by a central division at the base; blade often in a horizontal plane, but the youngest leaf usually with segments ± erect; spines on midrib above. [*Mauritia*]
7. b Segments many-ribbed, not oriented in a marked horizontal plane; spines present on the upper surface of the ribs and on blade edges (select the youngest leaves). [*Lepidocaryum*]
8. a Blade white beneath. [see 9]
8. b Blade green beneath. [see 10]
8. c Blade hazy glaucous beneath, with three to four leaflets, each usually with two bright green margins beneath. [*Oenocarpus*]
9. a Rib(s) prominent beneath; lower surface of the blade usually covered with a continuous layer of thin, white, membranous scales which rub off on contact; three to four segments not markedly erect. [*Chelyocarpus*, *Itaya*]
9. b Rib(s) not prominent beneath; blade dusty white beneath, petiole of the youngest leaves also dusty white; segments of the youngest leaf markedly erect. [*Mauritiella*]
10. a With triangular, laterally flattened spines on the petiole, these curved toward the apex as well as toward the base of the leaf. In the southern limit of the basin, Brazil and Bolivia. [*Copernicia*]

10. b Without spines on the edge of the petiole. [see 11]
11. a Fibers of the sheath joined in a spiny expansion. In the southern limit of the basin, Brazil and Bolivia. [*Trithrinax*]
11. b Fibers of the sheath not joined in a spiny expansion, or sheath not fibrous. [see 12]
12. a Midrib prominent above. [*Mauritia*]
12. b Midrib of each one-pointed segment prominent beneath. [*Chelyocarpus*]
13. a Leaflets or halves of bifid blade with tips pointed (Fig. 64c,d,e,f). [see 14]
13. b Leaflets or halves of bifid blade clearly truncate, denticulate, or multi-pointed (Fig. 64g,h,i). [see 30]
14. a Two leaflets connate at base, far longer than wide, or blade bifid. [see 15]
14. b Two leaflets not connate at base, wide, length about two to four times the width, usually light green, soft, not rigid. [*Desmoncus*]
15. a With spines, at times sparse, on outer edge of leaflets or halves of bifid blade, and, in some cases, on blade above and on sheath. [see 16]
15. b Without spines on edge of leaflets or halves of bifid blade, or on other leaf parts. [see 19]
16. a Blade white beneath. [*Astrocaryum*]
16. b Blade green or glaucous beneath. [see 17]
17. a Blade glaucous beneath, light green above; reddish, shiny spines on sheath, rachis, and blade. [*Acrocomia*]
17. b Blade green beneath. [see 18]
18. a Spines short, broad basally, up to 3–4 mm long, generally whitish, sometimes with brown tips, located on upper surface of ribs and on leaflet edges. [*Lepidocaryum*]
18. b Spines or spiny hairs, slender, never swollen at base, often dark, grouped on the outer edge near the tip of each half of bifid blade, occasionally on blade above; usually longer spines on sheath, petiole and rachis below, these brown or black, rarely whitish then flattened. [*Bactris*]
19. a Leaflets white beneath. [see 20]
19. b Leaflets or half of bifid blade green beneath. [see 22]
19. c Leaflets hazy glaucous with one or two bright green margins beneath. [*Oenocarpus*]
19. d Halves of bifid blade green beneath with grayish or brownish stripes along the ribs and the margins; outer margins denticulate toward the tip. [*Attalea*, *Maximiliana*, *Orbignya*, *Scheelea*]
20. a Two leaflets in closed-V shape, connate at base for 3 cm or more (Fig. 64c). [*Jessenia*]
20. b Two leaflets in open-V shape, connate at base for less than 3 cm long, each leaflet slightly twisted outwards (Fig. 64d). [see 21]
21. a Ribs prominent beneath; segments with layer of white membranous scales beneath which rub off on contact. [*Chelyocarpus*, *Itaya*]

21. b Ribs not prominent beneath; segments dusty white beneath; petiole usually dusty white. [*Mauritiella*]
22. a Petiole and rachis yellow, or very light green, or whitish; in cultivated areas. [*Cocos*]
22. b Petiole and rachis green. [see 23]
23. a Pinnae linear-lanceolate, or halves of bifid blade with margins straight (Fig. 64c,e). [see 24]
23. b Pinnae or halves of bifid blade S-shaped (Fig. 64f). [See 27]
24. a Pinnae or halves of bifid blade rather wide (more than 2 cm), one or several ribs. [see 25]
24. b Pinnae narrow (less than 2 cm), several ribs. [See 26]
25. a With two pinnae, and a distinct, long petiole between the sheath and the rachis. [*Euterpe*]
25. b With blade entire and bifid, and a short petiole, or without petiole, then the sheath continues into the rachis. [*Elaeis*]
26. a Ribs (three) prominent above and beneath; sheath tubular, shorter opposite the petiole (obliquely open). [*Wendlandiella*]
26. b Several ribs prominent above; sheath split opposite the petiole (select old, green leaves). [*Geonoma*]
27. a Sheath split opposite the petiole (select old, green leaves). [*Geonoma*, *Pholidostachys*, *Asterogyne*]
27. b Sheath not split opposite the petiole, tubular in young as well as in old leaves. [see 28]
28. a Ribs prominent above and beneath on pinnae. [see 29]
28. b Ribs prominent above on pinnae, not beneath. [*Hyospathe*]
29. a Outer margins of blade emarginate (or wavy). [*Chamaedorea*]
29. b Outer margins of blade regularly linear and curved, not emarginate. [*Wendlandiella*]
30. a Tip of pinnae or of halves of bifid blade regularly multi-pointed (Fig. 64g). [see 31]
30. b Tip of pinnae or of halves of bifid blade irregularly toothed, not regularly multi-pointed (Fig. 64h,i). [see 32]
31. a With brownish or grayish stripes on the blade beneath; blade length less than 1 m. [*Attalea*, *Maximiliana*, *Orbignya*, *Scheelea*]
31. b Without brownish or grayish stripes on the blade beneath; blade length up to several meters. [*Manicaria*]
32. a Pinnae one-ribbed, with the two margins straight; with spines on sheath, petiole, rachis, and/or blade. [*Aiphanes*]
32. b Pinnae or halves of bifid blade many-ribbed with the inner (or upper) margin straight, and the outer (or lower) margin wavy, irregularly toothed; without spines. [see 33]
33. a Blade divided for less than half its length (Fig. 64h); sheath, petiole, and blade pilose. [*Iriartella*]
33. b Blade divided for more than half its length (Fig. 64i); leaf not pilose. [see 34]

34. a Blade white beneath. [*Dictyocaryum*]
34. b Blade green beneath. [*Socratea*]
35. a Leaf length less than 2.5 m. [see 36]
35. b Leaf length more than 2.5 m, blade often torn along the nerves (leaf tip usually slightly bifid). *Manicaria*
36. a Blade far longer than wide, linear-lanceolate (Fig. 64j,k). [see 37]
36. b Blade round, or with length about two or three times the width (Fig. 64l,m). [see 40]
37. a Tip of blade pointed, prolonged by a narrow apex (Fig. 64j); petiole short or not distinct between sheath and blade. [*Elaeis*]
37. b Tip of blade not sharply pointed, or roundly blunt (Fig. 64k), or regularly denticulate. [see 38]
38. a Blade white beneath, usually with small, slender spines on edge, and on sheath and petiole. [*Astrocaryum*]
38. b Blade green beneath with longitudinal brownish or grayish stripes; without spines. [*Attalea*, *Maximiliana*, *Orbignya*, *Scheelea*]
38. c Blade green beneath, without brownish or grayish stripes; without spines. [see 39]
39. a Tip of blade roundly truncate, pilose; blade length less than 0.3 m. [*Iriartella*]
39. b Tip of blade neither rounded nor sharply pointed, not pilose, blade length up to 2.5 m. [*Syagrus*]
39. c Tip of blade regularly denticulate. [*Manicaria*]
40. a Tip of blade pointed (Fig. 64l). [see 41]
40. b Tip of blade not pointed (Fig. 64m). [see 42]
41. a Blade pilose beneath (not always obvious, then can be confused with *Wettinia*; difference can be made in identifying adult palms if present; stem no more than 5 cm in diameter, and leaf with less than 17 pairs of pinnae). [*Catoblastus*]
41. b Blade not pilose beneath (adult stem more than 6 cm diam., and leaf with more than 17 pairs of pinnae). [*Wettinia*]
42. a Blade, petiole and sheath strongly pilose. [*Iriartella*]
42. b Blade, petiole and sheath, not strongly pilose. [*Iriartea*]
43. a Pinnae pointed at tip, linear-lanceolate or S-shaped (Fig. 64,n,o,p). [see 44]
43. b Pinnae (or part when divided longitudinally) truncate at tip, lanceolate to wedge-shaped or like a fish fin (Fig. 64s,t,u). [see 70]
44. a Pinnae much longer than wide. [see 45]
44. b Pinnae wide, narrow basally, with length about two to four times the width, sub-opposite, two to four pairs per leaf; blade usually light green, soft, not rigid; rarely armed with small hooks or spines on the rachis and sheath. [*Desmoncus*]
45. a With spines on pinna edges. [see 46]
45. b Without spines on pinna edges. [see 48]
46. a Pinnae white beneath; spines strongly flattened, usually black, sometimes whitish. [*Astrocaryum*]

46. b Pinnae glaucous beneath; spines not strongly flattened, often reddish brown. [*Acrocomia*]
46. c Pinnae green beneath. [see 47]
47. a Spines or spiny hairs black or dark brown, slender, no more than 1 cm long, often shorter, on the edges of the pinnae, usually near the tip, or disposed more or less regularly; spines longer on sheath, petiole, and rachis (absent in some small species), sometimes on the blade above, dark not flattened, or whitish to yellowish with brown tip when strongly flattened. [*Bactris*]
47. b Spines whitish or brownish, 2–4 mm long, swollen basally, regularly disposed on the edges along the pinnae, and on the midrib above; sheath, petiole, and rachis unarmed. [*Raphia*]
48. a Pinnae lanceolate (Fig. 64n). [see 49] – If three or four ribs prominent above and beneath on a few narrow pinnae, two or three pairs per leaf. [see 69b]
48. b Pinnae S-shaped (Fig. 64o,p). [see 66]
49. a Rachis continuing into a cirrus with strong hooks in a V-shape; pinnae sub-opposite to opposite; sheath and petiole often with prickles basally swollen or with slender spines. [*Desmoncus*]
49. b Rachis not continuing into a cirrus with hooks. [see 50]
50. a Tip of pinnae symmetric on both sides on the midrib (select pinnae from several leaves; Fig. 64b). [see 51]
50. b Tips of pinnae asymmetric, obliquely notched, with acute or rounded tip (Fig. 64q,r). [see 59]
51. a Pinnae green beneath. [see 52]
51. b Pinnae white beneath (select the youngest leaves), wide (more than 6 cm), serrate in cross-section, normally arranged in one plane; sheath with erect, knitting-needlelike, black projections, 20 cm and more long. [*Jessenia*]
51. c Pinnae hazy glaucous beneath, generally less than 6 cm wide, normally oriented in the same plane, or in groups of two to six in different directions; the rachis and petiole of youngest leaves dusty red (the youngest leaf often with red blade); sheath fibrous at margins, sometimes forming a muff around the stem, often reddish brown, without knitting-needlelike projections. [*Oenocarpus*]
52. a With hooks on the petiole margins. [*Elaeis*]
52. b Without hooks on the petiole margins. [see 53]
53. a Pinnae arranged in one plane. [see 54]
53. b Pinnae oriented in several directions perpendicular to the rachis; leaves finely pinnate. [*Syagrus*]
54. a Pinnae obviously serrate (like a saw) in cross-section (see the basal parts); petiole triangular in cross-section; fibers at leaf bases. [*Barcella*]
54. b Pinnae not obviously serrate in cross-section; petiole round basally, not strongly triangular in cross-section. [see 55]

55. a Sheath with light brown fibers, split opposite the petiole; sheath, petiole, rachis, and midrib, light green, yellowish to whitish. [*Cocos*]
55. b Sheath with dense, black fibers, split opposite the petiole; sheath, petiole and rachis dark green. Pinnae sub-opposite forming an upward V-shape toward the tip and a downward V-shape toward the base. [see 56]
55. c Sheath without fibers, tubular, not split opposite the petiole, yellow, orange, or green. [see 58]
56. a Leaf sheath fibers flattened. [*Ammandra*]
56. b Leaf sheath fibers rounded or edged but never flattened [see 57]
57. a Apical pinnae alternate, rachis continuing into a short and narrow pinna. [*Phytelephas*]
57. b Apical pinnae opposite. [*Aphandra*]
58. a Pinna tip flat; sheath yellowish to orange in adult plants, usually covered by sheath remnants of dead leaves in juvenile plants. [*Euterpe*]
58. b Pinna tip carinate, sometimes flat in seedlings (confusion with *Euterpe* possible); sheath greenish. Genus limited to the western Amazonia and the Andean piedmont. [*Prestoea*]
59. a Pinnae green beneath. [see 60]
59. b Pinnae hazy glaucous beneath. [*Oenocarpus*]
59. c Pinnae green with longer tips brownish or grayish beneath (see pinnae of several leaves). [*Attalea*, *Maximiliana*, *Orbignya*, *Scheelea*]
59. d Pinnae white beneath. [*Jessenia*]
60. a With hooks on petiole margins. [*Elaeis*]
60. b Without hooks on petiole margins. [see 61]
61. a Pinnae oriented in several directions; pinna edges rough (like a small hack-saw); rachis often continuing into only one slender pinna. [*Syagrus*]
61. b Pinnae arranged in one plane; pinna edges smooth. [see 62]
62. a Midrib prominent on both upper and lower surfaces on pinnae, parallel nerves prominent beneath. Pinnae opposite forming an upward V-shape toward the tip and a downward V-shape toward the base. [see 63]
62. b Midrib prominent above, not beneath on pinnae. [see 65]
63. a Leaf sheath fibers flattened; apical pinnae opposite. [*Ammandra*]
63. b Leaf sheath fibers rounded or edged but never flattened. [see 64]
64. a Apical pinnae alternate, rachis continuing into a short and narrow pinna (this character is obvious; sometimes pinna tips are not strongly asymmetric). [*Phytelephas*]
64. b Apical pinnae opposite. [*Aphandra*]
65. a Pinna tip slightly bifid, sometimes symmetric, several nerves parallel to the midrib and more prominent on pinna above than beneath; sheaths of dead leaves persistent and forming a muff of fibers around the stem. [*Leopoldinia*]

65. b Pinna tip not bifid, carinate; several nerves parallel to the midrib and prominent beneath; without a persistent fibrous muff around the stem, this with well-marked internodes. [*Prestoea*]
66. a Leaf tip bifid with tip of each half one-pointed. [see 67]
66. b Leaf tip bifid with tip of each half multi-pointed. [*Manicaria*]
67. a Sheath split opposite the petiole (select old green leaves). [*Geonoma*, *Pholidostachys*]
67. b Sheath tubular, never split opposite the petiole. [see 68]
68. a Ribs prominent above and beneath on pinnae. [see 69]
68. b Ribs prominent above, not beneath on pinnae. [*Hyospathe*]
69. a Pinnae generally more than 2 cm wide, markedly S-shaped, with asymmetric tip, the lower part shorter than the upper; internodes like a truncate inverted cone; stem diameter more than 1 cm. [*Chamaedorea*]
69. b Pinnae narrow (no more than 2 cm wide), not markedly S-shaped, with symmetric tips, usually with three or four prominent ribs; two or three pairs of pinnae; internodes cylindrical; stem diameter less than 1 cm. [*Wendlandiella*]
70. a With spines on the sheath, petiole, rachis, and on the stem; pinna tip truncate and broad (Fig. 64s). [*Aiphanes*]
70. b Without spines; pinna wedge-shaped or like a fish fin (Fig. 64t,u). [see 71]
71. a With irritant hairs on the sheath. [*Iriartella*]
71. b Without irritant hairs on the sheath. [see 72]
72. a Pinnae undivided and arranged in one plane. [see 73]
72. b Pinnae longitudinally divided, parts oriented in several directions. [see 77]
73. a Leaf tip bifid. [see 74]
73. b Leaf tip not bifid. [see 75]
74. a Pinnae green beneath. [*Socratea*]
74. b Pinnae white beneath. [*Dictyocaryum*]
75. a Leaf tip pointed. [see 76]
75. b Leaf tip not pointed. [*Iriartea*]
76. a Blade pilose beneath (not always obvious, then could be confused with *Wettinia*; adult stem no more than 5 cm diam., and leaf with less than 17 pairs of pinnae). [*Catoblastus*]
76. b Blade not pilose beneath (adult stem more than 6 cm diam., and leaf with more than 17 pairs of pinnae). [*Wettinia*]
77. a Upper parts of divided pinnae slightly erect but with drooping tips (leaf viewed in cross-section, Fig. 64v), green beneath. [*Socratea*]
77. b Upper parts of divided pinnae straight and slightly erect (leaf viewed in cross-section). [see 78]
78. a Upper parts shorter than the lower (leaf viewed in cross-section, Fig. 64w), green beneath. [*Iriartea*]
78. b Upper parts not shorter than the lower (leaf viewed in cross-section, Fig. 64x), white beneath. [*Dictyocaryum*]

Key 2: Leaves Are Inaccessible (Palm Height More than 10–12 m) – Binoculars Will Be Helpful

1. a Leaves palmate (Fig. 63a) or costapalmate (Fig. 63b). [see 2]
1. b Leaves pinnate (Fig. 63e). [see 4]
2. a With spines on the trunk; most often multistemmed palms; leaf palmate. [*Mauritiella*]
2. b Without spines on the trunk; single-stemmed palms; leaf costapalmate or palmate. [see 3]
3. a Leaf costapalmate, without spines on the petiole. [*Mauritia*]
3. b Leaf palmate with triangular, laterally flattened spines on the petiole. [*Copernicia*]
4. a Pinnae linear-lanceolate with tip pointed. [see 5]
4. b Pinnae with truncate tip, narrow and long, slightly asymmetric to wedge-shaped. [see 20]
5. a Prickly palms (if no spines on the trunk, see sheath, petiole or rachis – for tall palms look at dead fallen leaves). [see 6]
5. b Unarmed palms. [see 9]
6. a Pinnae oriented in several directions (leaves ragged). [see 7]
6. b Pinnae regularly arranged in one plane. [*Astrocaryum*]
7. a Extremity of basal leaves lower than their point of insertion; with spines on the trunk, except in old palms, these not strongly flattened; upper pinnae slightly erect with drooping tips. [see 8]
7. b Extremity of basal leaves well above their point of insertion (crown funnel-like); spines on the trunk most often strongly flattened; pinnae straight, tips not drooping. [*Astrocaryum*]
8. a Trunk diameter more than 20 cm; numerous leaves radiating out to form a spherical crown. Several dead leaves pendent under the crown. [*Acrocomia*]
8. b Trunk diameter under 20 cm; basal leaves markedly arching. Dead leaves usually not pendent under the crown. [*Bactris*]
9. a Pinnae oriented in several planes or directions (ragged leaves). [see 10]
9. b Pinnae arranged in one plane. [see 14]
9. c Pinnae of each side drooping in two parallel, vertical planes; sheath yellowish to orange; often with red roots from trunk base, those bearing small, white, spiny roots. [*Euterpe*]
10. a Sheath and basal part of petiole of dead leaves persistent under the crown for more than 1 m. [see 11]
10. b Sheaths of dead leaves not persistent under the crown. [see 12]
11. a Leaves in crown arranged in four to six vertical series (see the vertical superposition of petioles from beneath). [*Maximiliana*]
11. b Leaves many, not arranged in obvious vertical series. [*Scheelea*]
12. a Sheath split opposite the petiole [see 13]
12. b Sheath not split opposite the petiole, tubular, tapered. [*Roystonea*]

13. a Leaf sheaths dark green, brown to reddish, fibrous, forming a prominent net below the crown. Pinnae glaucous beneath. [*Oenocarpus*]
13. b Leaf sheaths grayish, not forming a prominent net below the crown; pinnae green beneath, oriented in several directions perpendicular to the rachis making the leaf bottle-brushlike. [*Syagrus*]
14. a Sheath and petiole base of dead leaves persistent under the crown; leaves ascending to suberect, large (8–10 m long) with numerous pinnae (200 pairs and more), the tip of the leaf curving in the manner of a cock's tail feather. [*Attalea*, *Orbignya*, *Scheelea*]
14. b Sheath of dead leaves not persistent under the crown. [see 15]
15. a Large leaves, more than 4 m long; trunk diameter usually more than 20 cm. [see 16]
15. b Medium-sized leaves, less than 4 m long; trunk diameter less than 20 cm. [see 18]
16. a With hooks on the margin of the petiole. [*Elaeis*]
16. b Without hooks on the margin of the petiole. [see 17]
17. a Extremities of basal leaves above their point of insertion or at the same level, rarely below (crown open-funnel-like). Pinnae white beneath. [*Jessenia*]
17. b Extremities of basal leaves lower than their point of insertion; numerous leaves radiating out to form a spherical crown, often with a slightly free space between an upper and a lower group of leaves; sheath, petiole, rachis, and midrib yellowish to whitish. Pinnae green to yellowish beneath. [*Cocos*]
18. a Sheath tubular. [see 19]
18. b Sheath not tubular, split opposite the petiole, usually fibrous at margins. [*Oenocarpus*]
19. a Internodes well marked. [*Prestoea*]
19. b Internodes not distinctly marked. [*Euterpe*]
20. a Pinnae undivided, arranged in one plane; medium-sized palms. [see 21]
20. b Pinnae longitudinally divided to the base, parts oriented in several directions; tall palms. [see 22]
21. a Leaf tips entire (select the youngest leaves). [*Wettinia*]
21. b Leaf tips bifid. [*Socratea*]
22. a Upper parts of divided pinnae slightly erect with drooping tips (Fig. 64v), green beneath; numerous stilt roots regularly spaced forming a rather open cone up to 3 m in height, each stilt root light-brown bearing small, white, hornlike, sharp, spinelike roots; root cap small at the apex of growing roots. [*Socratea*]
22. b Upper parts of divided pinnae slightly erect with tips straight, not drooping. [see 23]
23. a Upper parts of divided pinnae shorter than the lower parts (Fig. 64w), green beneath; stilt roots at a very acute angle with the trunk forming

a rather closed cone up to 2 m in height, each stilt root dark-brown, bearing white, spinelike roots; large cap at the apex of growing roots. Stem usually with conspicuous swelling in low western Amazonia; stem usually without swelling in the Andean piedmont. [*Iriartea*]

23. b Upper parts of divided pinnae not shorter than the lower parts (Fig. 64x), white beneath; stilt roots at a rather obtuse angle with the trunk forming an open cone up to 1 m in height, each stilt root light brown, bearing white, spinelike roots. [*Dictyocaryum*]

2. Elementary Description of Amazonian Palm Genera

An elementary description of each genus, which takes into account only the Amazonian species, emphasizes the most conspicuous characters. The reader will find complete botanical descriptions with sumptuous illustrations in Uhl and Dransfield's *Genera Palmarum* (1987).

The geographic distribution is given according to three regions: (1) east – Atlantic coast to about Santarém (Brazil); (2) central – Santarém to about Tefé (Brazil), including Amazonian parts of Venezuela and Guianas; (3) west – Tefé to Andean piedmont, including Amazonian parts of Bolivia, Colombia, Ecuador, and Peru.

Fruit sizes are qualified by small (<3 cm), medium (3–6 cm), and large (>6 cm), considering the greatest dimension of the fruit.

Most common vernacular names used in Bolivia (B), Brazil (Br), Colombia (C), Ecuador (E), French Guiana (FG), Guyana (G), Peru (P), Surinam (S), and Venezuela (V), are given for each genus and then separately listed in alphabetical order with the corresponding species in the third part of this supplement. Data are from our field experience and from literature (Graham 1934; Simonson 1958; Wessels Boer 1965, 1988; Glassman 1972; Silva et al. 1977; Balick 1986; Balslev and Barfod 1987; Balslev and Moraes 1989; Borgtoft Pederson and Balslev 1990; Henderson 1990; Barfod 1991). Indian names, the uses of which are limited to the tribe territory, are excluded here.

1. *Acrocomia.* Large, single-stemmed palms, armed with slender spines on the trunk (this usually without spines in old palm), on leaf sheath, petiole, rachis, and pinnae. Crown physiognomy is characterized by being spherical with dead leaves hanging at the base. Inflorescence interfoliar with a long, erect peduncle. Infructescence pendent with many branches. Fruit medium to large, globose, smooth.
Distribution: native in northeastern and southern region of the basin; introduced in central Amazonia, near Manaus.
Ecology: in open vegetation on sandy soils, in coastal savannahs, and secondary forest.

Vernacular names: totai (B); coco de catarro, macaúba, mucajá (Br); grugru (Br, FG, V); mucaya, noix de coyol (FG); corozo de vino, palma de vino (V).

2. ***Aiphanes***. Small, single- or multistemmed, or subacaulescent, prickly palms. Leaves pinnate; pinnae blunt and broad at tip. Inflorescence interfoliar, spicate or branched, erect, usually with a long peduncle. Fruit small, globose, red, with fleshy pulp at maturity.
Distribution: in western Amazonia; infrequent in lowland forests, common on the Andean piedmont and on mountain slopes.
Ecology: in understory of wet forests.
Vernacular names: cocos rura (B); cuchi rabo (E): shica-shica (P); marará (V).

3. ***Ammandra***. Small, multistemmed palms with no more than two prostrate stems (Barfod 1991). Leaves pinnate; sheath with flattened fibers; pinnae regularly distributed in one plane. Inflorescence interfoliar. Staminate flowers grouped by six to nine on very short, flattened branches. Pistillate inflorescence headlike, unbranched. Fruit large, covered in large, pointed warts.
Distribution: in western Amazonia (Ecuador).
Ecology: in wet forests; abundant in pastures and deforested areas.
Vernacular name: yarina blanca (E).

4. ***Aphandra***. Medium-sized, single-stemmed, unarmed palms. Stem short, dark, roughly ringed, partially covered with black, slender, rounded, or edged fibers. Median pinnae regularly disposed in one plane, basal pinnae oriented in several directions, apical pinnae very short. Inflorescences interfoliar, the staminate pendent, with many flower clusters, the pistillate headlike, unbranched. Fruit large, covered in large, pointed warts.
Distribution: in western Amazonia (Ecuador and Peru).
Ecology: in wet forests; abundant in pastures and deforested areas.
Vernacular name: piassaba (E).

5. ***Asterogyne***. Small, subacaulescent or single-, short-stemmed, unarmed palms. Similar to *Geonoma*. Leaves entire, bifid. Inflorescence spicate. Fruit small, elliptic.
Distribution: only known from a locality in southern French Guiana.
Ecology: in seasonal swamp forest understory.
Vernacular name:?

6. ***Astrocaryum***. Large, medium-sized, or small, single- or multi-, or subterranean-stemmed palms; always armed with strongly flattened, usually black or dark brown, in some cases reddish, yellowish, or whitish spines, on stem, leaf sheath, petiole, rachis, and sometimes on pinnae. Several species

with unarmed, roughly ringed, dark trunk. Leaves pinnate; pinnae white beneath, oriented in several directions from the rachis or regularly disposed in one plane. Inflorescence interfoliar with a long, usually erect, rarely pendent peduncle. Bracts spiny, often fur-like. Fruit medium to large, several per branch, then smooth, globose to ellipsoid, greenish, yellowish, orange, or densely inserted on the rachis, then black-prickly, often densely congested and prismatic through the mutual pressure, brown to orange, mesocarp floury to fleshy; in some species, epicarp torn and open at maturity, looking like an orange-colored flower.
Distribution: throughout the Amazon basin.
Ecology: in all forest ecosystems and savannahs, on sandy, clayey, well-drained, waterlogged or flooded soils.
Vernacular names: chonta, chonta loro, chonta negra, chontilla (B); coqueiro javari, jauary, munbaca, murumuru, murumuru da terra firme, tucum assu, tucum bravo, tucum da serra, tucum do mato, tucum purupuru, tucumã, tucumã arara, tucumã-í, tucumã piririca, tucumã uassu-rana (Br); cumare, yauari, yavari (C); awarra, awarra liba, counana, jauari, mouroumourou (FG); akuyuro palm, awarra, awarra liba, cuyuru palm, sawarai, tucumou (G); chambira, huicungo, huiririrma (P); amana, boegroemaka, koenana, pakiramaka, paramaka, pingomaka, soela-awarra, toekoemau, warau (S); cumare, tucuma, yavaide (V).

7. *Attalea, Maximiliana, Orbignya, Scheelea.* These four related genera are essentially defined from the characters of the staminate flower. It is impossible to separate them clearly from only vegetative characters.

Large single- or subterranean-stemmed palms. Leaves pinnate, large; pinnae arranged in one plane (*Attalea*, *Orbignya*, *Scheelea*), or oriented in several directions (*Maximiliana*, *Scheelea*). Inflorescence interfoliar with a lignous, brownish bract, parallel-furrowed on the outside, pointed at apex, particularly large in *Maximiliana*. Infructescence erect (*Maximiliana*, and subterranean-stemmed species of *Orbignya* and *Scheelea*), or pendent (*Attalea*, *Orbignya* and *Scheelea*, sometimes in *Maximiliana*). Fruit medium to large in most species.
Distribution: throughout the Amazon basin.
Ecology: *Maximiliana* (only one species) occurs in terra firme forests and secondary vegetation of clayey soils, in savannahs on sandy soils; arborescent species of *Orbignya* are found in upland as well as in seasonal swamp forests; subterranean-stemmed species of *Orbignya* and *Scheelea* usually occur in understory of terra firme forests and on waterlogged, sandy soils; arborescent species of *Scheelea* and *Attalea* are common on periodically flooded alluvial soils.
Vernacular names: (1) *Attalea*: curua (Br); conta (P). (2) *Maximiliana*: motacusillo (B); inajá (Br); guichire (C); maripa (FG); kokerite palm (G); inayuca (P); cururito, yagua (V). (3) *Orbignya*: babassu, palha preta, palha vermelha, pindobé, sombreiro (Br); macoupi (FG); catirina (P); baboen

pina, bergi-maripa (S); mavaco (V). (4) *Scheelea*: motacú, palla (B); caiaté, jacy, palha branca, uacury, urucury (Br); canambo (E); macoupi (FG); contillo, shapaja, shebon (P).

8. *Bactris*. Small, multistemmed, sometimes subacaulescent, prickly palms. One cultivated species is taller (*Bactris gasipaes*). Spines black or brown, sometimes yellowish to whitish, usually abundant on leaf sheath and petiole, and on bract; the smallest species unarmed, except hairs on blade edge. Leaves entire and bifid, or with S-shaped or linear-lanceolate pinnae, arranged in one plane or oriented in several directions, regularly disposed or by groups. Inflorescence usually interfoliar or among the persistent sheaths of dead leaves, unbranched, bi- or trifurcate, or many-branched. Fruit usually small (large in the cultivated species), globose or ellipsoid, yellow, red or blue-black at maturity.
Distribution: throughout the Amazon basin.
Ecology: in understory of all forest ecosystems, in low vegetation on sandy soils, in flooded areas, in savannahs and secondary vegetation. Particularly diversified in understory of upland forests in central and western Amazonia.
Vernacular names: marayaú (B); marajá, marajá do igapó, marajazinha, piranga (Br); corozo de gallina (C); zagrinette (FG); ñejilla (P); bamba, heegronmaka, kaminamaka, kaw-maka, kikismaka (S); corocillo, corocito, cubarrito, cubarro, cubarro rebalsero, cucurito, espina, macanilla, moporo, uva montañera (V). Cultivated species: chonta fina (B); pupunha (Br); cachipay (C); pejibaye (C, V); chonta, chonta duro (E); parépou (FG); paripi palm, peach palm (G); pijuayo (P); paripoe (S); bobi, cachipaes, macanilla, pijiguao (V).

9. *Barcella*. Small, subacaulescent, unarmed palms. Leaves pinnate; petiole fibrous. Inflorescence erect with several, straight and thick branches. Fruit medium-sized, subglobose.
Distribution: genus of one species located in northern Amazonia, Roraima, Brazil.
Ecology: in low vegetation on sandy soils (Henderson 1986b).
Vernacular name: piassaba brava, piassaba rana (Br).

10. *Catoblastus*. Small, unarmed palms. Stem diameter 4–6 cm. Leaves imparipinnate. Pinnae like a fish fin. Inflorescence infrafoliar with several branches. Fruit small, longer than wide, covered in short brown hairs. Seedlings and acaulescent juveniles can be confounded with *Wettinia*.
Distribution: only one species of the genus occurs in the western region of the Amazon basin.
Ecology: in understory of forests on waterlogged soils.
Vernacular name: ponilla (P).

11. *Chamaedorea*. Small, clustered or single-stemmed, unarmed palms. Stem green, internode usually like a truncated inverted cone (this not

obvious in some species). Leaves bifid, or with linear-lanceolate or S-shaped pinnae. Inflorescence inter or infrafoliar, spicate or branched, often with orange flowers and branches. Fruit small, globose or oblong, usually black at maturity.
Distribution: in western Amazonia; frequent in the Andean piedmont and on mountain slopes.
Ecology: in understory of wet forests.
Vernacular names: tactilla (B); chontilla, deodorante de indígena (E); sangapilla (P).

12. *Chelyocarpus*. Small to medium, single or multiple, erect or prostrate-stemmed, unarmed palms. Leaves palmate, with multi-pointed segments, white beneath, or with one-pointed segments connate basally, green beneath. Inflorescence interfoliar, usually pendent, erect in the prostrate species. Bracts spoon-like, more or less carinate. Fruit small, brownish or greenish, smooth or tessellate.
Distribution: in western Amazonia.
Ecology: in understory of upland and periodically flooded forests on alluvial soils.
Vernacular names: hoja redonda (B); sacha aguajillo, sacha bombonaje (P).

13. *Cocos*. Tall, single-stemmed, unarmed palms. Leaves with pinnae arranged in one plane. Petiole, rachis, and midrib of the pinna usually light green, yellowish, or whitish. Inflorescence interfoliar, erect. Fruit very large.
Distribution: introduced in Amazonia.
Ecology: in gardens and fields where the species is cultivated; very characteristic of sandy beaches.
Vernacular names: coco, cocotero (B, C, E, P, V); coqueiro da baia (Br); coconut palm (G); cocotier (FG).

14. *Copernicia*. Tall, single-stemmed palms. Leaves palmate. Toothlike spines on petiole margins. Inflorescence interfoliar. Fruit small to medium.
Distribution: in the southern region adjacent to the Amazon basin, Brazil, and Bolivia.
Ecology: in dense populations in savannahs and dry forests.
Vernacular names: caranda negra, palma blanca, palma negra (B); carandá (Br).

15. *Desmoncus*. Small, climbing, prickly palms. Rachis of the leaf prolonged into a cirrus bearing V-shaped hooks, these often swollen basally. Pinnae usually sub-opposite, regularly disposed in one plane, spines or small hooks on sheath and petiole. Juvenile forms unarmed; rachis not prolonged into a cirrus. Inflorescence interfoliar, with a few to many branches. Fruit small, ellipsoid, red or brownish at maturity.
Distribution: throughout the Amazon basin.

Ecology: in understory of upland and swamp forests, in open vegetation.
Vernacular names: junco de buena vista; palma trepadora, palma uña de gato (B); titara (B, V); jacitará (Br, V); banga palm, kamawarri (G); cashapurina (P); bambamaka (S); albarico, camaure, jacitara, volador (V).

16. *Dictyocaryum.* Tall, single-stemmed palms with stilt roots. Leaves paripinnate. Pinnae longitudinally divided into segments oriented in several directions, the upper as long as the lower (leaf viewed in cross-section). Tips of pinnae usually straight not drooping. Blade white beneath. Inflorescence infrafoliar, pendent with many branches, those near the peduncle connate basally. Fruit medium, subglobose.
Distribution: in western Amazonia at low (130–200 m) and high (1500–2200 m) elevation.
Ecology: in upland forests on well-drained soils; also found on poorly drained soils in lower density.
Vernacular names: icho (B); bonbona paso (C); palma real (E); chonta, pona, pona colorada (P).

17. *Elaeis.* Large or medium-sized, single, erect or prostrate-stemmed palms. Leaves pinnate, large; pinnae regularly disposed in one or two planes at each side of the rachis, or more or less ragged; basal pinnae transformed into hooks. Staminate and pistillate flowers on separate inflorescences, both on the same plant, interfoliar, erect, with many short, thick branches. Fruit small to medium, ellipsoid, red or black when ripe.
Distribution: throughout the Amazon basin.
Ecology: the native species forms dense stands in seasonal swamp forests in western Amazonia (Kahn and Mejia 1986), it is also found on well-drained soils in Brazil, but said to be cultivated there (Ooi et al. 1981). The African species was introduced in many places, and can sometimes be found in secondary forest.
Vernacular names: palma aceitera africana, americana (B, C, E, P, V); caiaué, dendê, dendê do Pará (Br); corozo, corozo colorado (C, V); noli (C); awara-dende, palmier à huile (FG); African oil palm (G); poloponta (P); sabanna-obé (S).

18. *Euterpe.* Tall or medium-sized, single- or multistemmed, unarmed palms. Easy to identify by their pinnae usually drooping in two parallel planes, by the arching rachis, and the yellow to orange, tubular sheath. Inflorescence infrafoliar, rachis and branches obliquely erect, slightly pendent at maturity. Fruit small, globose, blue-black at maturity.
Distribution: throughout the Amazon basin.
Ecology: in seasonal swamp forests on waterlogged soils, in forests on periodically flooded alluvial soils; infrequent in upland forests. One species occurs on sandy, waterlogged soils.
Vernacular names: asaí, assai, palma de rosario (B); palmito (B, E, V); assaí, assaí chumbinho, assaí da mata, assaí do Pará, assaí mirim (Br);

huasai (E, P); palmito, pinot, wassaie (FG); manicola palm (G); baboen pina, monki-monki pina, pina pina, prasara (S); manaca, morroque, uassi (V).

19. *Geonoma*. Small, single- or multistemmed, or subacaulescent, unarmed palms. Stem usually slender, cane-like. Leaves entire, bifid, or with pinnae regularly or irregularly wide, arranged in one plane. Inflorescence inter or infrafoliar, spicate or branched. Flowers immersed in branches, these usually red at maturity. Fruit small, globose or ellipsoid, blue-black when ripe.
Distribution: throughout the Amazon basin.
Ecology: many species in the understory of terra firme forests, less diversified in wetland forest ecosystems and on mountain slopes where some species, however, can reach high densities.
Vernacular names: jatata (B); baruaru, ubim, ubim bravo, ubim juriti (Br); waï (FG); dahlibanni (G); palmiche (P); tas (S).

20. *Hyospathe*. Small, clustered, unarmed palms with slender stems. General aspect of *Geonoma*, but easy to distinguish by the tubular sheath never-split opposite the petiole. Inflorescence infrafoliar, peduncle and rachis slightly erect, branches straight, perpendicular to the rachis, yellowish in flowers, usually red at maturity. Fruit small, ellipsoid, globose, black when ripe.
Distribution: throughout the Amazon basin and on mountain slopes up to 1800–2000 m above sea level.
Ecology: in understory of upland and seasonal swamp forests.
Vernacular names: ubim rana (Br); palmiche (P); san pablo (V).

21. *Iriartea*. Tall, single-stemmed palms with a closed cone of dark-brown stilt roots basally. Sometimes with a swollen trunk. Leaves imparipinnate. Pinnae of adult palm longitudinally divided into slender, blunt segments oriented in several directions, the upper shorter than the lower (leaf viewed in cross-section); pinnae of juvenile palm entire, wedge-shaped, regularly disposed in one plane. Inflorescence infrafoliar, pendent with many branches, those near the peduncle connate basally. Fruit medium, globose, greenish to brownish.
Distribution: in western Amazonia, frequent in the Andean piedmont and on mountain slopes.
Ecology: in the canopy of wet forests.
Vernacular names: palma zancuda, pachuvilla (B); paxiuba barriguda (Br); cachuda barriguda (C); pambil (E); huacra pona (P).

22. *Iriartella*. Small, stoloniferous, unarmed palms, with hairs on the leaf sheaths, and on the dark-brown, slender stem. Pinnae like a fish fin. Inflorescence interfoliar, with several branches; infructescence infrafoliar. Fruit small, slightly reniform, brownish red at maturity.

Distribution: in central and western Amazonia.
Ecology: in the understory of upland forests on clayey to sandy, well-drained soils.
Vernacular names: paxiubinha, paxiubinha do macaco (Br); ponilla (P); cervatana (V).

23. ***Itaya.*** Medium-sized, single-stemmed palms. Leaves palmate with multi-pointed segments, white beneath. It will be distinguished in the field from *Chelyocarpus* by the petiole, which splits basally into the sheath. Fruit small, subglobose, epicarp smooth.
Distribution: only one species, in western Amazonia.
Ecology: in forests on periodically flooded alluvial soils.
Vernacular name: sacha bombonaje (P).

24. ***Jessenia.*** Large, single-stemmed palms. Leaves pinnate, large, erect; sheaths bearing long, knitting-needlelike, black projections; pinnae white beneath, regularly disposed in one plane. Inflorescence infrafoliar, horse tail-like, with many branches. Fruit small to medium, ellipsoid, blue when ripe.
Distribution: throughout the Amazon basin, less frequent in the south-eastern region.
Ecology: in seasonal swamp forests on permanently waterlogged soils and on sandy soils irregularly waterlogged after rainfall in central and western Amazonia; in terra firme forests on well-drained soils in Ecuador and French Guiana.
Vernacular names: majo, margarita (B); patauá (Br); bataua, comé, milpesos (C, E); chapil (E); patawa (FG); turu palm (G); ungurahui (P); patawa-koemboe (S); seje, seje grande (V).

25. ***Leopoldinia.*** Medium-sized, single- or multistemmed, unarmed palms. Usually with a fibrous muff around the stem made by the persistent sheaths of dead leaves. Leaves pinnate. Inflorescence interfoliar, with many sinuous branches. Fruit small to medium, more or less flat, lenticular.
Distribution: in central Amazonia, along the Rio Negro valley.
Ecology: in forests flooded by blackwater (igapó), occasionally in forests on alluvial soils.
Vernacular names: jará, jará-mirim, jará-assu, mucury, piassaba (Br); chique-chique (V).

26. ***Lepidocaryum.*** Small, multistemmed palms. Leaves palmate. Inflorescence interfoliar, pendent, with a few branches. Flowers yellowish. Fruit small, bearing scales, red at maturity.
Distribution: in western and central Amazonia.
Ecology: in upland forest understory.
Vernacular names: igaico (C); carana-í do matto (Br); irapay (P).

27. *Manicaria*. Medium-sized, single- or multistemmed, unarmed palms. Stem short, erect or leaning, conspicuously ringed with leaf scars. Leaves large, entire, more or less torn, sometimes irregularly pinnate. Infructescence interfoliar, pendent, with light brown, tissue-like bracts. Fruit medium to large one to three spheres thoroughly covered in warts.
Distribution: in patches throughout the northern region of the basin.
Ecology: in understory of seasonal swamp forests.
Vernacular names: bussu (Br); toulouri (FG); truli palm (G); toeroeli, troeli (S); temichi (V).

28. *Mauritia*. Large, single-stemmed palms with costapalmate leaves. Inflorescence interfoliar, pendent, branches arranged in one plane on each side of the rachis. Fruit medium to large, bearing scales, red at maturity.
Distribution: throughout the Amazon basin.
Ecology: in swamp forests and savannahs, particularly abundant on inundated acidic, organic soils. One species on waterlogged, white-sandy soils.
Vernacular names: palma real (B); burití, caraná, caraná grande, caranazinho, mirití, (Br); canangucha (C); aguaje (C, P); aete, ite palm (G); acho, aguaschi, morete (E); palmier bâche (FG); maurisie, morisi (S); moriche (V).

29. *Mauritiella*. Medium-sized to large, multistemmed palms. Stem armed with roots transformed into spines. Leaves palmate. Inflorescence interfoliar. Fruit medium, bearing scales, brownish red at maturity.
Distribution: throughout the Amazon basin.
Ecology: in seasonal swamp forests, in forests flooded by blackwater, in savannahs on waterlogged soils, in low vegetation on white-sandy soils.
Vernacular names: palmilla (B); buritirana, buritizinho, caraná, caraná de espinhos, carana-í (Br); moretillo (E); aguajillo (P); cahuáia, uliya (V).

30. *Maximiliana*. See *Attalea*.

31. *Oenocarpus*. Tall or medium-sized, single- or multistemmed, unarmed palms. Leaves pinnate, pinnae regularly disposed in one plane or by groups and oriented in several directions. Inflorescence infrafoliar, horse tail-like, pendent with many branches. Fruit small to medium, ellipsoid or ovoid, purplish black at maturity.
Distribution: throughout the Amazon basin.
Ecology: in terra firme forests, in swamp forests, in forests on periodically flooded alluvial soils.
Vernacular names: bacaba (B, Br); tarampabo (B); bacabinha, bacaba de leque (Br); manoco, milpesos, posuy, punama (C); milpesillo (C, E); comou (FG); turu palm (G); sinami, sinamillo (P); koemboe, koemoe (S): mapora, seje, sejito (V).

32. *Orbignya.* See *Attalea.*

33. *Pholidostachys.* Small, single-stemmed, unarmed palms. Very similar to *Geonoma*. Inflorescence interfoliar, branched. Fruit small, mesocarp with thick fibers (mesocarp is not fibrous in *Geonoma*).
Distribution: only one species of the genus in western Amazonia.
Ecology: in terra firme forest understory.
Vernacular name: palmiche grande (P).

34. *Phytelephas.* Medium-sized, single- or multistemmed, unarmed palms. Subacaulescent or with a dark, roughly ringed trunk. Graceful, pinnate leaves with fibrous sheath. Inflorescence interfoliar. Staminate flowers densely crowded on an unbranched axis, usually pendent. Pistillate flowers in clusters on an erect axis, between leaf bases. Infructescence headlike, unbranched. Fruit large, rounded in large, pointed warts.
Distribution: in western Amazonia.
Ecology: in understory of wet forests on mountain slopes; on alluvial soils.
Vernacular names: palma marfil (B, E); jariná (Br); yarina (Br, E, P); tagua (E); llarina, piasava (P).

35. *Prestoea.* Medium-sized to small, multistemmed or subacaulescent, unarmed palms. Leaves with pinnae arranged in one plane; sheath tubular. Inflorescence infrafoliar, slightly erect, usually with many branches. Fruit small, globose or subglobose, blue-black when ripe.
Distribution: in western Amazonia.
Ecology: in understory of wet forest in the Andean piedmont and on mountain slopes.
Vernacular names: palma ramosilla (B); palmito (E).

36. *Raphia.* Multistemmed palms with large, pinnate, ragged leaves. Pinnae with small, basally swollen spines regularly spaced on edges and on midrib above. Inflorescence apical, pendent. Fruit small to medium, ellipsoid, bearing scales.
Distribution: in eastern Amazonia.
Ecology: in seasonal swamp forests under tidal influence.
Vernacular names: jupaty, jurubaty (Br).

37. *Roystonea.* Tall, single-stemmed palm. Large trunk often tapered, swollen basally, gray or white, ringed by obscure leaf scars. Leaves pinnate, ragged; sheath tubular, tapered, forming a prominent crownshaft, green, Inflorescence infrafoliar, pendent. Fruit small, subglobose to ellipsoid.
Distribution: introduced in several regions of the basin as ornamental palm.
Ecology: in cities; found in a swamp forest in French Guiana (see Sect. 4.5.1.).
Vernacular names: palma real (C, P, V); palmier royal, palmiste (FG).

38. *Scheelea.* See *Attalea*.

39. *Socratea.* Tall or medium-sized, single-stemmed palms with stilt roots forming an open cone, up to 3(4) m in height; each stilt root light brown, bearing small, sharp, white, spine rootlets. Leaves paripinnate. Pinnae of the tall species longitudinally divided into slender segments oriented in several directions, the upper not clearly shorter than the lower; tips of pinnae drooping. Pinnae of the medium-sized species entire, asymmetric, like a fish fin, more or less wedge-shaped, regularly disposed in one plane. Inflorescence infrafoliar, pendent, with a few to many branches. Fruit medium, ovoid to ellipsoid, brown to yellowish.
Distribution: throughout the Amazon basin.
Ecology: the tall species occurs in all forest ecosystems, infrequent on sandy soils; the medium-sized species is found in the upper layers of terra firme forest understory in lowlands as well as on slopes of eastern Andean piedmont in Peru.
Vernacular names: pachiuva (B); pona (B, P), sancha pona (B); paxiuba (Br); bonbón, rayador (E); ponilla (E, P); awarra-mon-pè (FG); buba palm, stilt palm (G); casha pona (P); inga-pina, inga-prasara (S).

40. *Syagrus.* Medium-sized to large, single-stemmed or rarely multistemmed palms. Leaves of seedling entire, up to 2.5 m long. Leaves of stemmed juvenile and adult finely pinnate, ragged, sometimes entire; pinnae narrow and straight, oriented in several directions, usually perpendicular to the rachis. Inflorescence interfoliar, pendent, with several to many branches. Fruit medium to large, ellipsoid, yellow-orange, fleshy at maturity.
Distribution: throughout the Amazon basin.
Ecology: in terra firme forest understory, in wet forests in Andean piedmont valleys, and in dryer vegetation in the southern and northern regions of the basin.
Vernacular names: inaja-í, jará-rana, pupunha brava (Br); palmito (E); inshavi (P); peh-peh, pupunha de porco (S); sarare, zancona (V).

41. *Trithrinax.* Medium-sized palms with palmate leaves. Inflorescence interfoliar. Fruit small, globose, whitish.
Distribution: in the southern regions adjacent to the Amazon basin, Brazil, and Bolivia.
Ecology: in savannahs and gallery forests.
Vernacular names: palma chuco, saro (B).

42. *Wendlandiella.* Small, multistemmed, unarmed palms with slender cane. Pinnate leaves with tubular sheath. Pinnae usually with three prominent ribs. Inflorescence interfoliar. Fruit small.
Distribution: in western Amazonia.

Ecology: in understory of wet forest in Subandean region.
Vernacular name: ?

43. *Wettinia.* Medium-sized, single or multistemmed palms with stilt roots. Leaves imparipinnate; sheath tubular, green. Pinnae with shape of fish fin, regularly disposed in one plane (case of Amazonian species). Inflorescence infrafoliar, one or several at the node; peduncle branched or not. Fruit small, pilose, densely crowded on rachis, congested and prismatic through mutual pressure, gray brownish.
Distribution: in western Amazonia, common in the Andean piedmont and on mountain slopes.
Ecology: in upper layers of terra firme and seasonal swamp forest understory.
Vernacular names: gualte, gualte bola (E); cuyu-corota, ponilla (P).

3. Vernacular Names of Amazonian Palms

acho (E) – *Mauritia flexuosa*
aete (G) – *Mauritia flexuosa*
African oil palm (G) – *Elaeis guineensis*
aguaje (P) – *Mauritia flexuosa*
aguajillo (P) – *Mauritiella peruviana*
aguaschi (E) – *Mauritia flexuosa*
akuyuro palm (G) – *Astrocaryum aculeatum*
albarico (V) – *Desmoncus macroacanthos*
amana (S) – *Astrocaryum aculeatum*
asaí, assai (B) – *Euterpe precatoria*
assai (FG) – *Euterpe oleracea, E. precatoria*
assaí chumbinho (Br) – *Euterpe catinga*
assaí da mata (Br) – *Euterpe precatoria*
assaí do Pará (Br) – *Euterpe oleracea*
assaí mirim (Br) – *Euterpe precatoria*
awarra (FG, G) – *Astrocaryum vulgare*
awarra-dendé (FG) – *Elaeis guineensis*
awarra liba (FG, S) – *Astrocaryum jauari*
awarra-mon-pè (FG) – *Socratea exorrhiza*

babassu (Br) – *Orbignya phalerata*
baboen pina (S) – *Euterpe precatoria*
bacaba (Br) – *Oenocarpus bacaba*
bacaba (B) – *Oenocarpus* spp.
bacaba de leque (Br) – *Oenocarpus distichus*
bacabinha (Br) – *Oenocarpus minor*
bamba (S) – *Bactris simplicifrons*
bambamaka (S) – *Desmoncus orthacanthos, D. polyacanthos,*
banga palm (G) – *Desmoncus* sp.
baruaru (Br) – *Geonoma deversa*
bataua (C) – *Jessenia bataua* subsp. *bataua*
beach palm (G) – *Bactris major*
bergi-maripa (S) – *Orbignya sagotii*
bobi (V) – *Bactris gasipaes*
boegroemaka (S) – *Astrocaryum sciophilum*
bonbón (E) – *Socratea exorrhiza*
bonbona paso (C) – *Dictyocaryum ptariense*
buba palm (G) – *Socratea exorrhiza*
burití (Br) – *Mauritia flexuosa*
buritirana (Br) – *Mauritiella aculeata*
buritizinho (Br) – *Mauritiella aculeata*
bussu (Br) – *Manicaria martiana*

cachipaes (V) – *Bactris gasipaes*
cachipay (C) – *Bactris gasipaes*
cachuda barriguda (C) – *Iriartea deltoidea*
cahuaia (V) – *Mauritiella aculeata*
caiaté (Br) – *Scheelea* spp.
caiué (Br) – *Elaeis oleifera*
camaure (V) – *Desmoncus polyacanthos*
caña brava (C) – *Iriartella setigera*
canambo (E) – *Scheelea* sp.
canangucha (C) – *Mauritia flexuosa*
caraná (B, V) – *Mauritia carana*
carana (P) – *Lepidocaryum tessmannii*
caraná de espinhos (Br) – *Mauritiella aculeata*
caraná grande (Br) – *Mauritia carana*
carana-í (Br) – *Mauritiella aculeata*
carana-í do matto (Br) – *Lepidocaryum tenue*

caranazinho (Br) – *Mauritia carana*
carandá (Br) – *Copernicia alba*
caranda negra (B) – *Copernicia alba*
casha pona (P) – *Socratea exorrhiza*
cashapurina (P) – *Desmoncus* spp.
catirina (P) – *Orbignya polysticha*
cervatana (V) – *Iriartella setigera*
chambira (P) – *Astrocaryum chambira*
chapil (E) – *Jessenia bataua* subsp. *bataua*
chique chique (V) – *Leopoldinia piassaba*
chonta (B) – *Astrocaryum* spp.
chonta (E) – *Bactris gasipaes*
chonta duro (E) – *Bactris gasipaes*
chonta fina (B) – *Bactris gasipaes*
chonta loro (B) – *Astrocaryum ulei*
chonta negra (B) – *Astrocaryum* spp.
chontilla (B) – *Astrocaryum aculeatum*, *Astrocaryum vulgare*
chontilla, chontilla blanca (E) – *Chamaedorea pinnatifrons*
coco de catarro (Br) – *Acrocomia lasiospatha*
cocos rura (B) – *Aiphanes* sp.
cocotero (B, C, E, P, V) – *Cocos nucifera*
cocotier (FG) – *Cocos nucifera*
comé (C) – *Jessenia bataua* subsp. *bataua*
comou (FG) – *Oenocarpus bacaba*
conta (P) – *Attalea tessmannii*
contillo (P) – *Scheelea insignis*
coqueiro da baia (Br) – *Cocos nucifera*
coqueiro javari (Br) – *Astrocaryum jauari*
corocillo (V) – *Bactris oligoclada*
corocito (V) – *Bactris humilis*
corozo, corozo colorado (C, V) – *Elaeis oleifera*
corozo de gallina (C) – *Bactris major*
corozo de vino (V) – *Acrocomia lasiospatha*
counana (FG) – *Astrocaryum paramaca*
cubarrito (V) – *Bactris simplicifrons*
cubarro (V) – *Bactris bidentula*, *B. chaetochlamys*, *B. cruegeriana*, *B. major*, *B. maraja*
cubarro rebalsero (V) – *Bactris campestris*
cuchi rabo (E) – *Aiphanes* sp.
cucurito (V) – *Bactris cruegeriana*
cumare (C) – *Astrocaryum vulgare*
cumare (V) – *Astrocaryum aculeatum*, *A. chambira*, *A. vulgare*
cururito (V) – *Maximiliana maripa*
cuyu-corota (P) – *Wettinia maynensis*
cuyuru palm (G) – *Astrocaryum aculeatum*

dahlibanni (G) – *Geonoma baculifera*
dendê do Pará (Br) – *Elaeis oleifera*
deodorante de indigena (E) – *Chamaedorea pauciflora*

espina (V) – *Bactris integrifolia*

gru-gru (Br, FG, V) – *Acrocomia lasiospatha*
gualte (E) – *Wettinia maynensis*
gualte bola (E) – *Wettinia maynensis*
guara (C) – *Astrocaryum jauari*
guichire (C) – *Maximiliana maripa*

heegronmaka (S) – *Bactris elegans*
hoja redonda (B) – *Chelyocarpus chuco*
huacra pona (P) – *Iriartea deltoidea*
huasai (E, P) – *Euterpe precatoria*
huicungo (P) – *Astrocaryum macrocalyx*, *A. huicungo*
huiririrma (P) – *Astrocaryum jauari*

igaico (C) – *Lepidocaryum tessmannii*
inajá (Br) – *Maximiliana maripa*
inaja-í (Br) – *Syagrus inajai*
inayauca (P) – *Maximiliana maripa*
inga-pina (S) – *Socratea exorrhiza*
inga-prasara (S) – *Socratea exorrhiza*
inshavi (P) – *Syagrus sancona*
irapay (P) – *Lepidocaryum tessmannii*
ite palm (G) – *Mauritia flexuosa*

jacitará (Br) – *Desmoncus* spp.
jacitara (V) – *Desmoncus macroacanthos*
jacy (Br) – *Scheelea* spp.
jará (Br) – *Leopoldinia pulchra*
jará-assu (Br) – *Leopoldinia major*
jará-mirim (Br) – *Leopoldinia pulchra*
jará-rana (Br) – *Syagrus inajai*
jariná (Br) – *Phytelephas macrocarpa*
jatata (B) – *Geonoma deversa*, *G. jussieuana*
jauari (B, FG) – *Astrocaryum jauari*
junco de buena vista (B) – *Desmoncus* spp.
jupaty (Br) – *Raphia taedigera*
jurubaty (Br) – *Raphia taedigera*

kamawarri (G) – *Desmoncus* sp.
kaminamaka (S) – *Bactris gastoniana*
kaw-maka (S) – *Bactris major*
kikismaka (S) – *Bactris maraja*, *B. monticola*, *B. simplicifrons*
koemboe (S) – *Oenocarpus bacaba*
Koenana (S) – *Astrocaryum paramaca*
kokerite palm (G) – *Maximiliana maripa*

llarina (P) – *Phytelephas macrocarpa*

macanilla (V) – *Bactris gasipaes*, *B. major*
macaúba (Br) – *Acrocomia lasiospatha*
macoupi (FG) – *Orbignya* spp., *Scheelea* spp.

majo (B) – *Jessenia bataua* subsp. *bataua*
manaca (V) – *Euterpe oleracea*
manicola palm (G) – *Euterpe oleracea*
manoco (C) – *Oenocarpus bacaba*
mapora (V) – *Oenocarpus mapora*
maraja (Br) – *Bactris major*, *B. maraja*
maraja do igapó (Br) – *Bactris maraja*
marajazinha (Br) – *Bactris simplicifrons*
mararà (V) – *Aiphanes* sp.
marayaú (B) – *Bactris major*, *B.* sp.
margarita (B) – *Jessenia bataua* subsp. *bataua*
maripa (FG, S) – *Maximiliana maripa*
maurisie (S) – *Mauritia flexuosa*
mavaco (V) – *Orbignya polysticha*
milpesillo (C) – *Oenocarpus mapora*
milpesos (C, E) – *Jessenia bataua* subsp. *bataua*, *Oenocarpus mapora*
mirití (Br) – *Mauritia flexuosa*
monki-monki pina (S) – *Euterpe precatoria*
moporo (V) – *Bactris major*
morete (E) – *Mauritia flexuosa*
moretillo (E) – *Mauritiella aculeata*
moriche (C, V) – *Mauritia flexuosa*
morisi (S) – *Mauritia flexuosa*
morroque (V) – *Euterpe oleracea*
motacu (B) – *Scheelea princeps*, *S.* sp.
motacusillo (B) – *Maximiliana maripa*
moucaya (FG) – *Acrocomia lasiospatha*
mourou mourou (FG) – *Astrocaryum sciophilum*
mucajá (Br) – *Acrocomia lasiospatha*
mucury (Br) – *Leopoldinia pulchra*
munbaca (Br) – *Astrocaryum gynacanthum*
murumuru (Br) – *Astrocaryum macrocalyx*, *A. murumuru*
murumuru da terra firme (Br) – *Astrocaryum javarense*, *A. sociale*

ñejilla (P) – *Bactris* spp.
noix de coyol (FG) – *Acrocomia lasiospatha*
noli (C) – *Elaeis oleifera*

pachiuba (B) – *Socratea exorrhiza*
pachuvilla (B) – *Iriartea deltoidea*
pakiramaka (S) – *Astrocaryum gynacanthum*
palha (B) – *Orbignya phalerata*
palha branca (Br) – *Scheelea* sp.
palha preta (Br) – *Orbignya spectabilis*
palha vermelha (Br) – *Orbignya spectabilis*
palla (B) – *Scheelea* sp.
palm bong (S) – *Acrocomia lasiospatha*
palma blanca (B) – *Copernicia alba*
palma chuco (B) – *Trithrinax schizophylla*
palma de rosario (B) – *Euterpe precatoria*
palm de vino (V) – *Acrocomia lasiospatha*
palma marfil (B, E) – *Phytelephas macrocarpa*
palma negra (B) – *Copernicia alba*
palma ramosilla (B) – *Prestoea* spp.
palma real (B) – *Mauritia flexuosa*, *Orbignya phalerata*
palma real (C, P, V) – *Roystonea* spp.
palma trepadora (B) – *Desmoncus* spp.
palma uña de gato (B) – *Desmoncus polyacanthos*
palma zancuda (B) – *Iriartea deltoidea*
palmera aceitera africana (B, C, E, P, V) – *Elaeis guineensis*
palmera aceitera americana (B, C, E, P, V) – *Elaeis oleifera*
palmiche (P) – *Geonoma* spp., *Hyospathe elegans*
palmiche grande (P) – *Pholidostachys synanthera*
palmier à huile (FG) – *Elaeis guineensis*
palmier bâche (FG) – *Mauritia flexuosa*
palmier royal (FG) – *Roystonea oleracea*
palmilla (B) – *Mauritiella aculeata*
palmiste (FG) – *Roystonea oleracea*
palmito (E) – *Prestoea* spp., *Syagrus sancona*
palmito (B, FG) – *Euterpe precatoria*
palmito (V) – *Euterpe precatoria*
pambil (E) – *Iriartea deltoidea*
paramaka (S) – *Astrocaryum paramaca*
parépou (FG) – *Bactris gasipaes*
paripi palm (G) – *Bactris gasipaes*
paripoe (S) – *Bactris gasipaes*
patauá (Br) – *Jessenia bataua* subsp. *bataua*
patawa (FG, S) – *Jessenia bataua* subsp. *oligocarpa*
patawa-koemboe (S) – *Jessenia bataua* subsp. *oligocarpa*
paxiuba (Br) – *Socratea exorrhiza*
paxiuba barriguda (Br) – *Iriartea deltoidea*
paxiubinha (Br) – *Iriartella setigera*
paxiubinha do macaco (Br) – *Iriartella stenocarpa*
peach palm (G) – *Bactris gasipaes*
peh-peh (S) – *Syagrus inajai*
pejibaye (C, V) – *Bactris gasipaes*
piasava (P) – *Phytelephas macrocarpa*
piassaba (Br) – *Leopoldinia piassaba*
piassaba (E) – *Aphandra natalia*
piassaba brava (Br) – *Barcella odora*
piassaba rana (Br) – *Barcella odora*
pijiguao (V) – *Bactris gasipaes*
pijuayo (P) – *Bactris gasipaes*
pina (S) – *Euterpe oleracea*
pingomaka (S) – *Astrocaryum sciophilum*
pinot (FG) – *Euterpe oleracea*

piranga (Br) – *Bactris* spp.
poloponta (P) – *Elaeis oleifera*
pona (B) – *Socratea exorrhiza*
pona (P) – *Dictyocaryum lamarckianum*
pona colorada (P) – *Dictyocaryum ptariense*
ponilla (E, P) – *Catoblastus drudei*, *Iriartella stenocarpa*, *Socratea salazarii*, *Wettinia augusta*, *W. maynensis*
posuy (C) – *Oenocarpus mapora*
prasara (S) – *Euterpe oleracea*
punama (C) – *Oenocarpus bacaba*
pupunha (Br) – *Bactris gasipaes*
pupunha brava (Br) – *Syagrus inajai*
pupunha de porco (Br) – *Syagrus inajai*

rayador (E) – *Socratea exorrhiza*

sabanna-obé (S) – *Elaeis oleifera*
sacha aguajillo (P) – *Chelyocarpus repens*
sacha bombonaje (P) – *Chelyocarpus repens*, *C. ulei*, *Itaya amicorum*
sancha pona (B) – *Socratea exorrhiza*
sangapilla (P) – *Chamaedorea fragrans*, *C. pauciflora*
san pablo (V) – *Hyospathe elegans*
sarare (V) – *Syagrus sancona*
saro (B) – *Trithrinax schizophylla*
sawarai (G) – *Astrocaryum jauari*
seje (V) – *Jessenia bataua* subsp. *bataua*, *Oenocarpus bacaba*
seje grande (V) – *Jessenia bataua* subsp. *bataua*
sejito (V) – *Oenocarpus bacaba*
shapaja (P) – *Scheelea brachyclada*, *S. cephalotes*, *S. stenorhyncha*
shebon (P) – *Scheelea bassleriana*
shica shica (P) – *Aiphanes* spp.
sinami (P) – *Oenocarpus balickii*, *O. mapora*
sinamillo (P) – *Oenocarpus balickii*, *O. mapora*
soela-awarra (S) – *Astrocaryum jauari*
sombreiro (Br) – *Orbignya spectabilis*
stilt palm (G) – *Socratea exorrhiza*

tactilla (B) – *Chamaedorea lanceolata*
tagua (E) – *Phytelephas* spp.
tagua blanca (E) – *Ammandra dasyneura*
tarampabo (B) – *Oenocarpus tarampabo*
tas (S) – *Geonoma baculifera*
temichi (V) – *Manicaria atricha*
titara (B, V) – *Desmoncus macroacanthos*
toekoemau (S) – *Astrocaryum aculeatum*
toeroeli (S) – *Manicaria saccifera*
toulouri (FG) – *Manicaria saccifera*
totai (B) – *Acrocomia totai*
totai barrigudo (B) – *Acrocomia totai*
troeli (S) – *Manicaria saccifera*
truli palm (G) – *Manicaria saccifera*
tucum assu (Br) – *Astrocaryum aculeatum*
tucum bravo (Br) – *Astrocaryum vulgare*
tucum da serra (Br) – *Astrocaryum aculeatum*
tucum do matto (Br) – *Astrocaryum aculeatum*
tucum purupuru (Br) – *Astrocaryum aculeatum*
tucumã (Br) – *Astrocaryum aculeatum*
tucuma (V) – *Astrocarum aculeatum*
tucumã arara (Br) – *Astrocaryum aculeatum*
tucumã piririca (Br) – *Astrocaryum aculeatum*
tucumã uassu rana (Br) – *Astrocaryum aculeatum*
tucumã-í (Br) – *Astrocaryum acaule*
tucumou (G) – *Astrocaryum aculeatum*
turu palm (G) – *Jessenia bataua* subsp. *oligocarpa*, *Oenocarpus bacaba*

uacury (Br) – *Scheelea princeps*
uassi (V) – *Euterpe oleracea*
ubim (Br) – *Geonoma* spp.
ubim bravo (Br) – *Geonoma deversa*
ubim juriti (Br) – *Geonoma deversa*
ubim rana (Br) – *Hyospathe elegans*
uliya (V) – *Mauritiella aculeata*
ungurahui (P) – *Jessenia bataua* subsp. *bataua*
urucuri (Br) – *Scheelea* sp.
uva montañera (V) – *Bactris monticola*

volador (V) – *Desmoncus orthacanthos*

waï (FG) – *Geonoma* spp.
warau (S) – *Astrocaryum aculeatum*
wassaïe (FG) – *Euterpe oleracea*

yagua (V) – *Maximiliana maripa*
yarina (P) – *Phytelephas macrocarpa*
yarina blanca (E) – *Ammandra dasyneura*
yauari, yavari (C) – *Astrocaryum jauari*
yavaide (V) – *Astrocaryum aculeatum*

zagrinette (FG) – *Bactris* spp.
zancona (V) – *Syagrus sancona*

References

Adis J (1984) "Seasonal igapó"-forests of Central Amazonian blackwater rivers and their terrestrial arthropod fauna. In: Sioli H (ed) The Amazon limnology and landscape ecology of a mighty tropical river and its basin. Dr Junk, Dordrecht, pp 245–268

Altman RRA, Cordeiro MMCM (1964) A industrialização do fruto do burití (*Mauritia vinifera* Mart. ou *M. flexuosa*). INPA Publ 5 Química, Manaus

Amedegnato C, Poulain S (1986) Diagnoses et signalisations des Romaleidae arboricoles amazoniens (Orthoptera Acrididae). Ann Soc Entomol Fr NS 22:423–455

Anderson AB (1978) The names and uses of palms among a tribe of Yanomama indians. Principes 22:30–41

Anderson AB (1981) White-sand vegetation of Brazilian Amazonia. Biotropica 13:199–210

Anderson AB (1983) The biology of *Orbignya martiana* (Palmae), a tropical dry forest dominant in Brazil. Thesis, University of Florida, Gainesville

Anderson AB (1988) Use and management of native forests dominated by açaí palm (*Euterpe oleracea* Mart.) in the Amazon estuary. Adv Econ Bot 6:144–154

Anderson AB, Anderson S (1985) A "tree of life" grows in Brazil. Nat Hist 94:41–46

Anderson AB, Prance GT, Albuquerque BWP de (1975) Estudos sobre a vegetação das campinas amazônicas. III. A vegetação lenhosa da campina da Reserva Biológica INPA-SUFRAMA (Manaus-Caracaraí, Km 62). Acta Amazonica 5:225–246

Anderson AB, Gély A, Strudwick J, Sobel GL, Pinto MGC (1985) Um sistema agroflorestal na várzea do estuário amazônico (Ilha das Onças, Município de Barcarena, Estado do Pará). Acta Amazonica 15 (Suppl):195–224

Anderson AB, Overal WL, Henderson A (1988) Pollination ecology of a forest-dominant palm (*Orbignya phalerata* Mart.) in northern Brazil. Biotropica 20:192–205

Anderson AB, May PH, Balick MJ (1991) The subsidy from nature. Columbia Univ Press, New York

Anonymous (1975) Inventario, evaluación e integración de los recursos naturales de la zona de Iquitos, Nauta, Requena y Colonia Angamos. Oficina Nacional de Evaluación de Recursos Naturales, Lima

Anonymous (1977) Use of remote sensing systems evaluating the potential of the aguaje palm tree in the Peruvian jungle. Oficina Nacional de Evaluación de Recursos Naturales, Lima

Anonymous (1983) Palmeras poco utilizadas de America tropical. FAO-CATIE, Turrialba, pp 113–117

Ashton PS (1969) Speciation among tropical forest trees: some deductions in the light of recent evidence. Biol J Linn Soc 1:155–196

Aubréville A (1961) Etude écologique des principales formations forestières du Brésil. Centre Technique Forestier Tropical, Nogent-sur-Marne

Balée W (1988) Indigenous adaptation to Amazonian palm forests. Principes 32:47–54

Balick MJ (1979a) Amazonian oil palms of promise. A survey. Econ Bot 33:11–28

Balick MJ (1979b) Economic botany of the Guahibo I. Palmae. Econ Bot 33:361–376

Balick MJ (1981) *Jessenia bataua* and *Oenocarpus* species: native Amazonian palms as new sources of edible oil. In: Pryde EH, Princen LH, Mukherjee KD (eds) New sources of fats and oils. American Oil Chemists' Society, Champaign, pp 141–155

Balick MJ (1984) Ethnobotany of palms in the neotropics. Adv Econ Bot 1:9–23
Balick MJ (1985a) Useful plants of Amazonia: a resource of global importance. In: Prance GT, Lovejoy TE (eds) Key environments: Amazonia. Pergamon Press, Oxford, pp 339–368
Balick MJ (1985b) The indigenous palm flora of "Las Gaviotas," Colombia, including observations on local names and uses. Bot Mus Leafl Harv Univ 30:1–34
Balick MJ (1986) Systematics and economic botany of the *Oenocarpus-Jessenia* (Palmae) complex. Adv Econ Bot 3:1–140
Balick MJ (1988) The use of palms by the Apinayé and Guajajara Indians of northeastern Brazil. Adv Econ Bot 6:65–90
Balick MJ, Anderson AB (1986/1987) Dry matter allocation in *Jessenia bataua* (Palmae). Acta Amazonica 16/17:135–139
Balick MJ, Gershoff SN (1981) Nutritional evaluation of the *Jessenia bataua* palm: source of high quality protein and oil from Tropical America. Econ Bot 35:261–271
Balslev H (1987) Palmas nativas de la Amazonía ecuatoriana. Colibrí 3:64–73
Balslev H, Barfod A (1987) Ecuadorean palms – an overview. Opera Bot 92:17–35
Balslev H, Henderson A (1986) *Elaeis oleifera* (Palmae) encontrada en el Ecuador. Publ Mus Cienc Nat Ecuador 5:45–49
Balslev H, Moraes M (1989) Sinopsis de las palmeras de Bolivia. AAU Rep 20, Aarhus
Balslev H, Luteyn J, Ollgaard B, Holm-Nielsen LB (1987) Composition and structure of adjacent unflooded and floodplain forest in Amazonian Ecuador. Opera Bot 92:37–57
Barbosa Rodrigues J (1903): Sertum palmarum brasiliensium. Imprimerie Monnom, Bruxelles, 2 vols
Barfod AS (1989) The rise and fall of vegetable ivory. Principes 33:181–190
Barfod AS (1991) A monographic study of the subfamily Phytelephantoideae (Arecaceae). Opera Bot 105:1–73
Barfod AS, Henderson A, Balslev H (1987) A note on the pollination of *Phytelephas microcarpa* (Palmae). Biotropica 19:191–192
Barfod AS, Bergmann B, Pedersen HB (1990) The vegetable ivory industry: surviving and doing well in Ecuador. Econ Bot 44:293–300
Beach JH (1984) The reproductive biology of the peach or "Pejibayé" palm (*Bactris gasipaes*) and a wild congener (*B. porschiana*) in the Atlantic lowlands of Costa Rica. Principes 28:107–119
Blaak G (1983) Procesamiento de los frutos de la palmera cucurita (*Maximiliana maripa*). In: Palmeras poco utilizadas de America tropical. FAO-CATIE, Turrialba, pp 113–117
Black GA, Dobzhansky T, Pavan C (1950) Some attempts to estimate species diversity and population density of trees in Amazonian forests. Bot Gaz 111:413–425
Blancaneaux P, Pouyllau M (1977) Formes d'altérations pseudokarstiques en relation avec la géomorphologie des granites précambriens du type Rapakivi dans le territoire fédéral de l'Amazone, Vénézuéla. Cah ORSTOM Sér Pédol XV(2):131–142
Bodley JH, Benson FC (1980) Stilt-root walking by an irarteoid palm in the Peruvian Amazon. Biotropica 12:67–71
Bondar G (1936) Notas biológicas sobre Bruchídeos observados no Brasil. Arquivos do Instituto Biológico de Vegetação (Rio de Janeiro Brasil) 3:1–44
Bondar G (1940) Notas entomológicas da Bahia VI. Rev Entomol 11:842–862
Bondar G (1941) Notas entomológicas da Bahia VII. Rev Entomol 12:268–303
Bondar G (1942) Notas entomológicas da Bahia X. Rev Entomol 13:225–274
Bondar G (1943) Notas entomológicas da Bahia XI. Rev Entomol 13:337–388
Bondar G (1964) Palmeiras do Brasil. Inst Bot São Paulo 2:1–159
Bongers F, Engelen D, Klinge H (1985) Phytomass structure of natural plant communities on spodosols in southern Venezuela: the Bana woodland. Vegetatio 63:13–34
Boom BM (1986) A forest inventory in Amazonian Bolivia. Biotropica 18:287–294
Boom BM (1988) The Chácobo Indians and their palms. Adv Econ Bot 6:91–97
Borgtoft Pederson H, Balslev H (1990): Ecuadorean palms for agroforestry. AAU Rep 23, AAU Univ Press, Aarhus

Borowiec L (1987) The genera of seed-beetles (Coleoptera, Bruchidae). Bull Entomol Pol 57:3–207

Bouillenne R (1924) Les racines-échasses de *Iriartea exorrhiza* Mart. (Palmiers) et de *Pandanus* div. sp. (Pandanacées). Mém Acad R Belg Cl Sci 8:1–45

Bouillenne R (1930) Un voyage botanique dans le bas Amazone. Arch Inst Bot Univ Liège 8:1–185, pls 1–34

Boulet R, Brugière JM, Humbel FX (1979) Relation entre organisation des sols et dynamique de l'eau en Guyane française septentrionale: conséquence agronomique d'une évolution déterminée par un déséquilibre d'origine principalement tectonique. Sci sol, Bulletin de l'AFES 1:3–18

Braga PIS (1977) Aspectos biológicos das Orchidaceae de uma campina da Amazônia Central. Acta Amazonica 7 (Suppl):1–89

Brokaw NVL (1982a) The definition of treefall gap and its effect on measures of forest dynamics. Biotropica 14:158–160

Brokaw NVL (1982b) Tree falls: frequency, timing and consequences. In: Leigh EG Jr, Rand AS, Windsor DM (eds) The ecology of a tropical forest: seasonal rhythms and long-term changes. Smithsonian Institution Press. Washington, DC, pp 101–108

Brokaw NVL (1985a) Treefalls, regrowth, and community structure in tropical forests. In: Pickett STA, White PS (eds) The ecology of natural disturbance and patch dynamics. Academic Press, New York, pp 53–69

Brokaw NVL (1985b) Gap-phase regeneration in a tropical forest. Ecology 66:682–687

Burret M (1934) Die Palmengattung *Astrocaryum* G. F. W. Meyer. Rep Spec Nov Regni Veg 35:114–158

Calzavara BBG (1972) As possibilidades do açaizeiro no estuário amazônico. Bol Fund Ciên Agrá Pará FCAP 5:1–103

Campbell DG, Daly DC, Prance GT, Maciel UN (1986) Quantitative ecological inventory of terra firme and várzea tropical forest on the Rio Xingú, Brazilian Amazon. Brittonia 38:369–393

Cavalcante PB (1974) Frutas comestíveis da Amazônia. Publicacões avulsas Museo Paraense Emílio Goeldi 27 Tome II, Belém

Charles-Dominique P (1986) Inter-relations between frugivorous vertebrates and pioneer plants: *Cecropia*, birds and bats in French Guiana. In: Estrada A, Fleming TH (eds) Frugivorous and seed dispersal, Dr Junk, Dordrecht, pp 119–135

Charles-Dominique P, Atramentowicz M, Charles-Dominique M, Gerard H, Hladik A, Hladik CM, Prevost MF (1981) Les mammifères frugivores arboricoles nocturnes d'une forêt guyanaise: inter-relations plantes animaux. Rev Ecol (Terre et Vie) 35:341–435

Chauvel A (1982) Os latossolos amarelos, alicos, argilosos dentro dos ecosistemas das bacias experimentais do INPA e da região vizinha. Acta Amazonica 12 (Suppl):47–60

Chauvel A, Lucas Y, Boulet R (1987) On the genesis of the soil mantle of the region of Manaus, central Amazonia, Brazil. Experientia 43:234–241

Chazdon RL (1984) Ecophysiology and architecture of three rain forest understory palm species. Thesis, Cornell Univ, Ithaca

Chazdon RL (1986a) Light variation and carbon gain in rain forest understory palm. J Ecol 74:995–1012

Chazdon RL (1986b) Physiological and morphological basis of shade tolerance in rain forest understory palms. Principes 30:92–99

Clement CR (1988) Domestication of the pejibaye palm (*Bactris gasipaes*): past and present. Adv Econ Bot 6:155–174

Clement CR, Mora Urpi JE (1987) Pejibaye palm (*Bactris gasipaes*, Arecaceae): multi-use potential for the lowland humid tropics. Econ Bot 41:302–311

Cooper A (1982) A phytosociological survey of the intermediate savannas of Guyana. Carib J Sci 17:81–92

Coradin L, Lleras E (1983) Situación actual de la investigación y desarrollo en palmeras poco conocidas. Informes por país. 1. Brasil. In: Palmeras poco utilizadas de América tropical. FAO-CATIE, Turrialba, pp 28–33

Coradin L, Lleras E (1988) Overview of palm domestication in Latin America. Adv Econ Bot 6:175–189
Corner EJH (1966) The natural history of palms. Weidenfeld and Nicolson, London
Couturier G (1988) Nuevas observaciones sobre los insectos de las palmeras nativas y los depredadores de los cultivos en la Amazonía peruana. ORSTOM/IIAP Rep, Lima
Couturier G, Kahn F (1989) Bugs of *Lincus* spp. vectors of Marchitez and Hartrot (oil palm and coconut diseases) on *Astrocaryum* spp., Amazonian native palms. Principes 33:19–20
Daly DC, Prance GT (1989) Brazilian Amazon. In: Campbell DG, Hammond HD (eds) Floristic inventory of tropical countries. New York Botanical Garden, New York, pp 401–426
Davis TA (1984) A climbing device to reach the palmyra's valuable "Toddy". Spirit of Enterprise. The 1984 Rolex Awards. Arum Press, London
Davis TAW, Richards PW (1933) The vegetation of Moraballi Creek, British Guiana: an ecological study of a limited area of tropical rain forest. I. J Ecol 21:350–384
de Foresta H, Charles-Dominique P, Erard C, Prevost MF (1984) Zoochorie et premiers stades de la régénération naturelle après coupe en forêt guyanaise. Rev Ecol (Terre et Vie) 39:369–409
de Granville JJ (1974) Aperçu sur la structure des pneumatophores de deux espèces des sols hydromorphes en Guyane *Mauritia flexuosa* L. et *Euterpe oleracea* Mart. (Palmae). Généralisation au système respiratoire racinaire d'autres palmiers. Cah ORSTOM Sér Biol 23:3–22
de Granville JJ (1976) Notes guyanaises: quelques forêts sur le Grand Inini. Cah ORSTOM Sér Biol XI:23–34
de Granville JJ (1977) Notes biologiques sur quelques palmiers guyanais. Cah ORSTOM Sér Biol XII:347–353
de Granville JJ (1978) Recherches sur la flore et la végétation guyanaise. Thesis, USTL II, Montpellier
de Granville JJ (1984) Monocotyledons and pteridophytes, indicators of environmental constraints in the tropical vegetation. Candollea 39:265–269
de Granville JJ (1985) Cueillir sans détruire. Poster, SEPANGUY, Cayenne
de Granville JJ (1986) Les formation végétales de la bande côtière de Guyane française. In: Le littoral guyanais. Fragilité de l'environnement. Nature Guyanaise n° hors série, Xème Colloque SEPANRIT, Cayenne 1985, pp 47–63
de Granville JJ (1988) Phytogeographical characteristics of the Guianan forests. Taxon 37:578–594
de Granville JJ (1989) La distribución de las palmas en la Guyana francesa. Acta Amazonica 19:115–138
de Granville JJ (1990) Les palmiers de la Guyane française. Bois For Trop 220:43–54
de Granville JJ, Henderson A (1988) A new species of *Asterogyne* (Palmae) from French Guiana. Brittonia 40:76–80
De Los Heros GMJ, Zárate JB (1980/1981) Posibilidades papeleras de pulpa al sulfato de pecíolos de aguaje. Rev For Peru 10:83–90
De Steven D (1986) Comparative demography of a clonal palm (*Oenocarpus mapora* subsp. *mapora*) in Panama. Principes 30:100–104
De Steven D, Windsor DM, Putz FE, Leon B de (1987) Vegetative and reproductive phenologies of a palm assemblage in Panama. Biotropica 19:342–356
Denslow JS (1980) Gap partitioning among tropical rainforest trees. Biotropica 12 (Suppl):47–55
Denslow JS, Newell E, Ellison AM (1991) The effect of understory palms and cyclanths on the growth and survival of *Inga* seedlings. Biotropica 23:225–234
Descamps M (1984) Revue préliminaire de la tribu des Copiocerini (Orth. Acrididae). Mém Mus Natn Hist Nat Sér A Zool 130:1–72
Desmier de Chenon R (1984) Recherches sur le genre *Lincus* Stal, Hemiptera Pentatomidae Discocephalinae, et son rôle éventuel dans la transmission de la Marchitez du palmier à huile et du Hart-Rot du cocotier. Oléagineux 39:1–6

Desmier de Chenon R, Merland E, Genty P, Morin JP, Dollet M (1983) Research on the genus *Lincus*, Pentatomidae Discocephalinae, and its possible role in the transmission of the Marchitez of oil palm and Hart-Rot of coconut. IV Reun del Com Rec Reg Sanidad Vegetal SARH-IICA, Cancún, Mexico

Dransfield J (1978) Growth forms of rain forest palms. In: Tomlinson PB, Zimmermann MH (eds) Tropical trees as living systems. Cambridge Univ Press, Cambridge, pp 247–268

Dransfield J, Uhl NW (1986). An outline of a classification of palms. Principes 30:163–167

Ducke A, Black GA (1953) Phytogeographical notes on the Brazilian Amazon. Anais Acad Bras Ciên 25:1–46

Dugand A (1972) Las palmeras y el hombre. Cespedesia 1:31–103

Encarnación F (1985) Introducción a la flora y vegetación de la Amazonía peruana: estadio actual de los estudios, medio natural y ensayo de una clave de determinación de las formaciones vegetales en la llanura Amazónica. Candollea 40:237–252

Ernould M (1921) Recherches anatomiques et physiologiques sur les racines respiratoires. Mem Acad R Belg Cl Sci Sér 2 6:1–52

Fanshawe D (1952) The vegetation of British Guiana: a preliminary review. Inst Pap 29, Commonw for Inst, Oxford

FAO-UNESCO (1971) Soil map of the world. Vol IV, South America. UNESCO, Paris

Fearnside PM (1982) Deforestation in the Brazilian Amazon: how fast is it occurring? Interciência 7:82–88

Fearnside PM (1984) A floresta pode acabar? Cienç Hoje 2:42–52

Fearnside PM, Rankin JM (1979) Avaliação da Jari Florestal e Agropecuária Ltda, como modelo para o desenvolvimento da Amazonia. Acta Amazonica 9:609–615

Fearnside PM, Rankin JM (1982) Jari and Carajás: the uncertain future of large silvicultural plantations in the Amazon. Interciência 7:326

Fittkau EJ, Klinge H (1973) On biomass and trophic structure of the Central Amazonian rain forest. Biotropica 5:2–14

Franken M, Irmler U, Klinge H (1979) Litterfall in inundation, riverine and terra firme forest of Central Amazonia. Trop Ecol 20:225–235

Galeano G (1991) Las palmas de la región de Araracuara. Tropenbos-Colombia, Bogotá

Gasc JP (1986) Le peuplement herpétologique d'*Astrocaryum paramaca* (Arécacées), un palmier important dans la structure de la forêt en Guyane française. Mém Mus Natn Hist Nat Sér A 132:97–107

Gatin GL (1907) Sur le développement des pneumathodes de palmiers et la véritable nature de ces organes. CR Acad Sci 144:649–651

Gentry AH (1982) Patterns of neotropical plant species diversity. Evol Biol 15:1–84

Gentry AH (1985) Algunos resultados preliminares de estudios botánicos en el Parque Nacional del Manu. In: Reporte Manu. Universidad Nacional Agraria, Lima, pp 1–23

Gentry AH (1986a) Sumario de patrones fitogeográficos neotropicales y sus implicaciones para el desarrollo de la Amazonía. Rev Acad Colomb Cienc Exactas Fis Nat Vol XVI 61:101–116

Gentry AH (1986b) Notes on Peruvian palms. Ann M Bot Gard 73:158–165

Gentry AH, López Parodi J (1980) Deforestation and increased flooding of the upper Amazon. Science 210:1354–1356

Glassman SF (1972) A revision of B. E. Dahlgren's index of American palms. J Cramer, Lehre

Gómez-Pompa A, Vázquez-Yanes C, Guevara S (1972) The tropical rain forest: a nonrenewable resource. Science 177:762–765

Gonzáles Rivadeneyra M (1971) Estudio sobre la densidad de poblaciones de Aguaje (*Mauritia* sp.) en Tingo María, Perú. Rev For Perú 5:41–53

Gottsberger G (1978) Seed dispersal by fish in the inundated regions of Humaitá, Amazônia. Biotropica 10:170–183

Goulding M (1980) The fishes and the forest. Univ of California Press, Berkeley

Graham EH (1934) II. Flora of the Kartabo Region, British Guiana. Ann Carnegie Mus 22:17–292

Greenhall AP (1965) Sapucaia nut dispersal by greater spear-nosed bats in Trinidad. Carib J Sci 5:167–171

Grenand P, Moretti C, Jacquemin H (1987) Pharmacopées traditionnelles en Guyane. Mém 108, ORSTOM, Paris

Guillaumet JL (1967) Recherche sur la végétation et la flore du Bas-Cavally. Mém 20, ORSTOM, Paris

Guillaumet JL (1973) Une nouvelle espèce de palmier nain de Madagascar. Adansonia Sér 2 13:341–349

Guillaumet JL (1987) Some structural and floristic aspects of the forest. Experientia 43:241–251

Guillotin M (1981) Données écologiques sur les petits rongeurs forestiers terrestres de Guyane française. Thesis, USTL II, Montpellier

Haffer J (1969) Speciation in Amazonian forest birds. Science 165:131–137

Hallé F, Oldeman RAA (1970) Essai sur l'architecture et la dynamique de croissance des arbres tropicaux. Masson et cie, Paris

Hallé F, Oldeman RAA, Tomlinson PB (1978) Tropical trees and forests: an architectural analysis. Springer, Berlin Heidelberg New York

Hartley CWS (1977) The oil palm, 2nd edn. Longmans, London

Hartshorn GS (1978) Tree falls and tropical forest dynamics. In: Tomlinson PB, Zimmermann MH (eds) Tropical trees as living systems. Cambridge Univ Press, Cambridge, pp 617–638

Hartshorn GS (1980) Neotropical forest dynamics. Biotropica 12 (Suppl):23–30

Heaney LR, Thorington RW (1978) The ecology of neotropical red-tailed squirrel *Sciurus granatensis* in the Panama canal zone. J Mammal 59:846–851

Heinen HD, Ruddle K (1974) Ecology, ritual and economic organization in the distribution of palm starch among the Warao of the Orinoco delta. J Anthropol Res 30:116–138

Henderson A (1985) Pollination of *Socratea exorrhizà* and *Iriartea ventricosa*. Principes 29:64–71

Henderson A (1986a) A review of pollination studies in the Palmae. Bot Rev 52:221–259

Henderson A (1986b) *Barcella odora*. Principes 30:74–76

Henderson A (1990) Arecaceae. Part I. Introduction and the Iriarteinae. Flora Neotropica, Monograph 53, New York

Henderson A, Balick MJ (1987) Notes of the palms of Amazônia Legal. Principes 31:116–122

Herrera R (1977) Soil and terrain conditions in the international Amazon project at San Carlos de Rio Negro, Venezuela, correlation with vegetation types. Transactions of the Int MAB-IUFRO Worksh on Tropical rainforest ecosystem research. In: Brunig EF (ed) Proc Symp, 12–17 May 1977, Hamburg-Reinbek, pp 193–199 (Chair World For Spec Rep, vol 1)

Herrera R (1985) Nutrient cycling in Amazonian forests. In: Prance GT, Lovejoy TE (eds) Key environments: Amazonia. Pergamon Press, Oxford, pp 95–108

Herrera R, Jordan CF, Klinge H, Medina E (1978) Amazon ecosystems: their structure and functioning with particular emphasis on nutrients. Interciencia 3:223–232

Heyligers PC (1963). Vegetation and soil of a white sand savanna in Suriname. In: Lanjouw J Versteegh PJD (eds) The vegetation of Suriname III. Van Eedenfonds, Amsterdam

Hogan KP (1986) Plant architecture and population ecology in the palms *Socratea durissima* and *Scheelea zonensis* on Barro Colorado Island, Panama. Principes 30:105–107

Holthuijzen AMA, Boerboom JHA (1982) The Cecropia seed bank in the Suriname lowland forest. Biotropica 14:62–68

Holttum RE (1955) Growth habits of monocotyledons – variations on a theme. Phytomorphology 5:399–413

Hoock J (1971) Les savanes guyanaises: Kourou. Mém 44, ORSTOM, Paris
Houghton RA, Lefkowitz DS, Skole DL (1991) Changes in the landscape of Latin America between 1850 and 1985 I. Progressive loss of forests. For Ecol Manage 38:143–172
Huber J (1906) Les palmiers du rio Purus. Bull Herb Bois 2nd Sér VI:266–271
Huber J (1910) Mattas e madeiras amazônicas. Bot Mus Paraense Nat Hist 6:91–225
Huttel C (1977) Etude de quelques caractéristiques structurales de la végétation du bassin versant de l'Audrenisrou (Côte d'Ivoire). ORSTOM Report, Abidjan
Irmler U (1977) Inundation-forest types in the vicinity of Manaus. Biogeographica 8:17–29
Irmler U, Furch K (1980) Weight, energy, and nutrient changes during the decomposition of leaves in the emersion phase of Central-Amazonian inundation forests. Pedobiologia 20:118–130
Izawa K (1979) Foods and feeding behaviour of wild black-capped capuchin (*Cebus apella*). Primates 20:57–76
Izawa K, Mizuno A (1977) Palm-fruit cracking behaviour of wild black-capped capuchin (*Cebus apella*). Primates 18:773–792
Janzen DH (1971) The fate of *Scheelea rostrata* fruits beneath the parent tree: predispersal attack by bruchids. Principes 15:89–101
Janzen DH (1978) A bat generated by seed shadow in rain forest. Biotropica 10:78
Janzen DH, Miler Ga, Hackforth-Jones J, Pond CM, Hooper K, Janos DP (1976) Two Costa Rica bat-generated seed shadow of *Andira inermis* (Leguminosae). Ecology 57:1068–1075
Jeník J (1973) Root system of tropical trees. 8. Stilt roots and allied adaptations. Preslia 45:250–264
Johson D (1975) Some palm products of the Peruvian Amazon. Principes 19:78–79
Johnson D (1982) Commercial palm products of Brazil. Principes 26:141–143
Johnson D (1988) Worldwide endangerment of useful palms. Adv Econ Bot 6:268–273
Jordan CB (1970) A study of germination and use in twelve palms of northeastern Peru. Principes 14:26–32
Jordan CF (1978) Stem flow and nutrient transfer in the tropical rain forest. Oikos 31:257–263
Jordan CF (1985) Soils of the Amazon rainforest. In: Prance GT, Lovejoy TE (eds) Key environments Amazonia. Pergamon Press, Oxford, pp 83–94
Jordan CF, Heuveldop J (1981) The water budget of an Amazonian rain forest. Acta Amazonica 11:87–92
Jordan CF, Golley FB, Hall J (1980) Nutrient scavenging of rainfall by the canopy of an Amazonian rain forest. Biotropica 12:61–66
Jost L (1887) Ein Beitrag zur Kenntniss der Athmungsorgane der Pflanzen. Bot Zeit 37:601–606, 617–627, 633–642
Journaux A (1975) Recherches géomorphologiques en Amazonie brésilienne. Bull 20, CNRS Centre de géomorphologie, Caen
Junk WJ (1984) Ecology of the várzea, floodplain of Amazonian whitewater rivers. In: Sioli H (ed) The Amazon limnology and landscape ecology of a mighty tropical river and its basin. Dr Junk, Dordrecht, pp 215–243
Junk WJ, Furch K (1985) The physical and chemical properties of Amazonian waters and theirs relationships with the biota. In: Prance GT, Lovejoy TE (eds) Key environments Amazonia. Pergamon Press, Oxford, pp 3–17
Kahn F (1977) Analyse structurale des systèmes racinaires des plantes ligneuses de la forêt tropicale dense humide. Candollea 32:321–358
Kahn F (1982) La reconstitution de la forêt tropicale humide (Sud-Ouest de la Côte d'Ivoire). Mém 97, ORSTOM, Paris
Kahn F (1983) Architecture comparée de forêts tropicales humides et dynamique de la rhizosphère. Thesis. USTL II, Montpellier
Kahn F (1986a) Les palmiers des forêts tropicales humides du bas Tocantins (Amazonie brésilienne). Rev Ecol (Terre et Vie) 41:3–13

Kahn F (1986b) Adaptation en forêt tropicale humide: le cas des palmiers amazoniens. Naturalia monspeliensia HS l'Arbre: 171–189
Kahn F (1986c) Life forms of Amazonian palms in relation to forest structure and dynamics. Biotropica 18:214–218
Kahn F (1987) The distribution of palms as a function of local topography in Amazonian terra firme forests. Experientia 43:251–259
Kahn F (1988a) A distichous *Mauritia flexuosa*. Principes 32:88
Kahn F (1988b) Ecology of economically important palms in Peruvian Amazonia. Adv Econ Bot 6:42–49
Kahn F (1990a) Identification of Amazonian palm genera from vegetative characters. Principes 34:199–207
Kahn F (1990b) Clave para diferenciar los géneros de Palmae en la Amazonía a partir del aparato vegetativo complementada por una breve descripción de los géneros, la distribución geográfica y ecológica, y los nombres vernáculos más comunes. Bull Inst Fr Etudes Andines 19:351–378
Kahn F (1991) Palm as key swamp forest resources in Amazonia. For Ecol Manage 38:133–142
Kahn F, Castro A (1985) The palm community in a forest of central Amazonia, Brazil. Biotropica 17:210–216
Kahn F, Mejia K (1986) The American oil palm, *Elaeis oleifera*, in Peruvian Amazonia. Principes 30:182
Kahn F, Mejia K (1987) Notes on the biology, ecology and use of a small Amazonian palm: *Lepidocaryum tessmannii*. Principes 31:14–19
Kahn F, Mejia K (1988) A new species of *Chelyocarpus* (Palmae, Coryphoideae) from Peruvian Amazonia. Principes 32:69–72
Kahn F, Mejia K (1990) Palm communities in wetland forest ecosystems of Peruvian Amazonia. For Ecol Manage 33/34:169–179
Kahn F, Mejia K (1991) The palm communities of two terra firme forests in Peruvian Amazonia. Principes 35:22–26
Kahn F, Mejia K, Castro A (1988) Species richness and density of palms in terra firme forests of Amazonia. Biotropica 20:266–269
Kitzke ED, Johnson D (1975) Commercial palm products other than oils. Principes 19:3–25
Klinge H (1965) Podzol soils in the Amazon Basin. J Soil Sci:96–103
Klinge H (1977a) Preliminary data on nutrient release from decomposing leaf litter in a neotropical rain forest. Amazoniana 6:196–202
Klinge H (1977b) Fine litter production and nutrient return to the soil in the tree natural forest stands in eastern Amazonia. Geol Ecol Trop:159–167
Klinge H (1978) Studies on the ecology of Amazon Caatinga in southern Venezuela. 2 Acta Cient Venez 29:258–262
Klinge H, Herrera R (1983) Phytomass structure of natural plant communities on spodosols in southern Venezuela: the tall Amazon caatinga forest. Vegetatio 53:65–84
Klinge H, Medina E (1979) Rio Negro caatingas and campinas, Amazon States of Venezuela and Brazil. In: Specht RL (ed) Heathlands and related shrublands. Descriptive studies. Elsevier, Amsterdam, pp 483–488 (Ecosystems of the World, vol 9A)
Klinge H, Rodrigues WA (1971) Matéria orgânica e nutrientes na mata de terra firme perto de Manaus. Acta Amazonica 1:69–72
Klinge H, Medina E, Herrera R (1977) Studies on the ecology of Amazon Caatinga forest in southern Venezuela. 1 Acta Cient Venez 28:270–276
Klinge H, Junk WJ, Revilla CJ (1990) Status and distribution of forested wetlands in tropical South America. For Ecol Manage 33/34:81–101
Korytkowski GCA, Ruiz AER (1979) El barreno de los racimos de la palma aceitera, *Castnia daedalus* (Cramer), Lepidopt.: Castniidae, en la plantación de Tocache, Perú. Rev Peru Entomol 22:49–62
Kramer F (1933) De natuurlijke verjonging in het Goenoeng-Gedehcomplex. Tectonia 26:156–185

Lamotte S (1990) Fluvial dynamics and succession in the lower Ucayali basin, Peruvian Amazonia. For Ecol Manage 33/34:141–156

Lechthaler R (1956) Inventário das árvores de um hectare de terra firme da Reserva Florestal Ducke, Município de Manaus, Amazônia. Rio de Janeiro Bot 3:1–10

Leopoldo PR, Franken W, Matsui E, Salati E (1982a) Estimativa de evapotranspiração de floresta amazônica de terra-firme. Acta Amazonica 12 (Suppl):23–28

Leopoldo PR, Franken W, Salati E (1982b) Balanço hídrico de pequena bacia hidrográfica em floresta amazônica de terra-firme. Acta Amazonica 12:333–337

Leopoldo PR, Franken W, Salati E, Ribeiro MNG (1987) Towards a water balance in the Central Amazonian region. Experientia 43:222–233

Lepesme P (1947) Les insectes des palmiers. Lechevallier, Paris

Lescure JP, Boulet R (1985) Relationships between soil and vegetation in a tropical rain forest in French Guiana. Biotropica 17:155–164

Lévi-Strauss C (1952) The use of wild plants in tropical South America. Econ Bot 6:252–270

Lindeman JC, Moolenaar SP (1959) Preliminary survey of the vegetation types of northern Suriname. In: de Hulster IA, Lanjouw J (eds) The vegetation of Suriname, vol 1, part 2. Van Eedenfonds, Amsterdam

Lisbôa RCL (1976) Estudos sobre a vegetação das campinas. V. Brioecologia. Acta Amazonica 6:171–192

Listabarth C (1992) A survey of pollination strategies in the Bactridinae. Bull Inst Fr Etudes Andines (in press)

Lleras E, Coradin L (1983) Anexo 7. La palmera macaúba (*Acrocomia aculeata*) como fuente potencial de aceite combustible. In: Palmeras poco utilizadas de America tropical. FAO-CATIE, Turrialba, pp 102–112

Lleras E, Coradin J (1988) Native neotropical oil palms: state of the art and perspective for Latin America. Adv Econ Bot 6:201–213

Llosa JF, Couturier G, Kahn F (1990) Notes of the ecology of *Lincus spurcus* and *L. malevolus* (Heteroptera: Pentatomidae: Discocephalinae) on Palmae in forests of Peruvian Amazonia. Ann Soc Entomol Fr NS 26:249–254

López Parodi J (1988) The use of palms and other native plants in non-conventional, low cost rural housing in the Peruvian Amazon. Adv Econ Bot 6:119–129

López Parodi J, Freitas D (1990) Geographical aspects of forested wetlands in the lower Ucayali, Peruvian Amazonia. For Ecol Manage 33/34:157–168

Louise C, Dollet M, Mariau D (1986) Recherches sur le Hartrot du cocotier, maladie à *Phytomonas* (Trypanosomatidae) et sur son vecteur *Lincus* sp. (Pentatomidae) en Guyane. Oléagineux 41:437–446

Luizão FJ, Schubart HOR (1987) Litter production and decomposition in a terra-firme forest of Central Amazonia. Experientia 43:259–265

Marmillod D (1982) Methodik und Ergebnisse von Untersuchungen über Zusammensetzung und Aufbau eines Terrassenwaldes im peruanischen Amazonien. Thesis, Georg-August Univ, Göttingen

Martínez Ramos M (1980) Aspectos sinecológicos del proceso de renovación natural de una selva alta perennifolia. Thesis, UNAM, Mexico

May PH, Anderson AB, Balick MJ, Frazão JMF (1985a) Subsistence benefits from the babassu palm (*Orbignya martiana*). Econ Bot 39:113–129

May PH, Anderson AB, Frazão JMF, Balick MJ (1985b) Babassu palm in the agroforestry systems in Brazil's mid-north region. Agrofor Syst 3:275–295

Medeiros-Costa JT, Panizza S (1983) Palms of the cerrado vegetation formation of São Paulo State, Brazil. Principes 27:118–125

Mejia K (1983) Palmeras y el selvicola amazónico. UNMSM, Mus Hist Nat Lima, pp 1–6

Mejia K (1988) Utilization of palms in eleven Mestizo villages of the Peruvian Amazon (Ucayali River, Department of Loreto). Adv Econ Bot 6:130–136

Meunier J (1976) Les prospections de palmacées. Une nécessité pour l'amélioration des palmiers oléagineux. Oléagineux 31:153–157

Milton K (1980) The foraging strategy of howler monkeys. Columbia Univ Press, New York

Moldenke HN (1949) The botanical source of patava oil. Phytologia V:122–129
Moore HE Jr (1973) Palms in the tropical forest ecosystems of Africa and South America. In: Meggers BJ, Ayensu ES, Duckworth DW (eds) Tropical forest ecosystems in Africa and South America. A comparative review. Smithsonian Institution Press, Washington, pp 63–88
Mora-Urpi J (1983) El pejibaye (*Bactris gasipaes* H.B.K.): Origen, biología floral y manejo agronómico. In: Palmeras poco utilizadas de América Tropical. FAO-CATIE, Turrialba, pp 118–160
Moran E (1984) Amazon basin colonization. Interciência 9:377–385
Mori SA, Boom BM (1987) Chapter II. The forest. In: The Lecythidaceae of a lowland neotropical forest: La Fumée mountain. Mem NY Bot Gard 44, New York, pp 9–29
Nierstraz E (1975) Clairière et chablis en forêt naturelle. ORSTOM Report, Abidjan.
Nordin CF Jr, Meade RH (1982) Deforestation and increased flooding of the upper Amazon (comments). Science 215:426–427
Oldeman RAA (1969) Etude biologique des pinotières de la Guyane française. Cah ORSTOM Sér Biol 10:1–18
Oldeman RAA (1971) Un *Eschweilera* (Lecythidaceae) à pneumatophores en Guyane française. Cah ORSTOM Sér Biol 15:21–27
Oldeman RAA (1974) L'architecture de la forêt guyanaise. Mém 73, ORSTOM, Paris
Oldeman RAA (1978) Architecture and energy exchange of dicotyledonous trees in the forest. In: Tomlinson PB, Zimmermann MH (eds) Tropical trees as living systems, Cambridge Univ Press, Cambridge, pp 535–560
Oldeman RAA (1983) Tropical rain forest, architecture, sylvigenesis and diversity. In: Sutton SL, Whitmore TC, Chadwick AC (eds) Tropical rain forest: ecology and management. Blackwell, Oxford, pp 139–150
Oldeman RAA (1990) Forests: elements of silvology. Springer, Berlin Heidelberg New York
Olesen JM, Balslev H (1990) Flower biology and pollinators of the Amazonian monoecious palm, *Geonoma macrostachys*: a case of Bakerian mimicry. Principes 34:181–190
Ooi SC, Da Silva FB, Müller AA, Nascimiento JC (1981) Oil palm genetic resources–native *Elaeis oleifera* populations in Brazil offer promising sources. Pesqũi Agropecũ Bras 16:385–395
Otedoh MO (1977) The African origin of *Raphia taedigera* – Palmae. Nigerian Field 42:11–16
Padoch C (1988) Aguaje (*Mauritia flexuosa* L.f.) in the economy of Iquitos, Peru. Adv Econ Bot 6:214–224
Perthuis B, Desmier de Chenon R, Merland E (1985) Mise en évidence du vecteur de la Marchitez sorpresiva du palmier à huile, la punaise *Lincus lethifer* Dolling (Hemiptera Pentatomidae Discocephalinae). Oléagineux 40:473–476
Pesce C (1985) Oil palms and others oil seeds of the Amazon. Translated and cited by DV Johnson from the original, Oleaginosas da Amazonia 1941. Oficina Graficas da Revista da Veterinaria, Belém. Reference Publ, Algonac, Michigan
Peters CM, Balick MJ, Kahn F, Anderson AB (1989) Oligarchic forests of economic plants in Amazonia. Environ Conserv 3:341–349
Piedade MTF (1984) Ecológia e biológia reproductiva de *Astrocaryum jauari* Mart. (Palmae) como exemplo de população adaptada ás áreas inundáveis do Rio Negro (igapós). Thesis, INPA/FUA, Manaus
Piñero D, Sarukhán J (1982) Reproductive behaviour and its individual variability in a tropical palm, *Astrocaryum mexicanum*. J Ecol 70:461–472
Piñero D, Martínez-Ramos M, Mendoza A, Alvarez-Buylla E, Sarukhán J (1986) Demographic studies in *Astrocaryum mexicanum* and their use in understanding community dynamics. Principes 30:108–116
Pinto GP (1951) O óleo de Patauá. Bol Tec IAN Belém 23:67–77
Pires JM (1978) The forest ecosystems of the Brasilian Amazon: description, functioning and research needs. In: Tropical forest ecosystems. UNESCO/UNEP/FAO, Paris, pp 607–627

Pires JM, Prance GT (1985) The vegetation types of the Brazilian Amazon. In: Prance GT, Lovejoy TE (eds) Key environments Amazonia. Pergamon Press, Oxford, pp 109–145

Pires JM, Dobzhansky T, Black GA (1953) An estimate of the number of species of trees in an Amazonian forest community. Bot Gaz 114:467–477

Porto ML, Longhi HM, Citadini V, Ramos RJ, Mariath JE de A (1976) Levantamento fitossociológico em área de "mata-de-baixio" na Estação Experimental de Silvicultura Tropical. INPA, Manaus. Acta Amazonica 6:301–318

Prance GT (1973) Phytogeographic support for the theory of Pleistocene refuges in the Amazon basin, based on evidence from distribution patterns in Caryocaraceae, Chrysobalanaceae, Dichapetalaceae and Lecythidaceae. Acta Amazonica 3:5–28

Prance GT (1978) The origin and evolution on the Amazon flora. Interciência 3:207–222

Prance GT (1979) Notes on the vegetation of Amazonia III. The terminology of Amazonian forest types subject to inundation. Brittonia 31:26–38

Prance GT (1982a) A review of the phytogeographic evidences for Pleistocene climate changes in the neotropics. Ann MO Bot Gard 69:594–624

Prance GT (1982b) Biological diversification in the tropic. Columbia Univ Press, New York

Prance GT (1986) The conservation and utilization of the Amazon rain forest. Rev Acad Colomb Cienc Exactas Fis Nat XVI(61):117–127

Prance GT, Lovejoy TE (eds) (1985) Key environments Amazonia. Pergamon Press, Oxford

Prance GT, Schubart HOR (1978) Notes on the vegetation of Amazonia. I. A preliminary note on the origin of the open white sand campinas of the lower Rio Negro. Brittonia 30:60–63

Prance GT, Rodrigues WA, Silva MF da (1976) Inventário florestal de um hectare de mata de terra firme km 30 da Estrada Manaus-Itacoatiara. Acta Amazonica 6: 9–35

Puig H, Delobelle JP (1988) Production de litière, nécromasse, apports minéraux au sol par la litière en forêt guyanaise. Rev Ecol (Terre Vie) 43:3–22

Putz FE (1979) Biology and human use of *Leopoldinia piassaba*. Principes 23:149–156

Ranzani G (1980) Identificação e caracterizacão de alguns solos da Estação Experimental de Silvicultura Tropical do INPA. Acta Amazonica 10:7–41

Ranzani G, Podesta JA (1982) Projeto Tucuruí. Relatório anual, Convenio INPA/ Eletronorte, Manaus

Rawitscher F, Rachid M (1946) Troncos subterrâneos de plantas brasileiras. Anais Acad Bras Ciên 18:261–280

Revilla JC (1981) Aspectos floristicos e fitossocioecológicos da floresta inudável (igapó) Praia Grande, Rio Negro, Amazônas, Brasil. Thesis, INPA-FUA, Manaus

Ribeiro MNG (1976) Aspectos climatológicos de Manaus. Acta Amazonica 6(2):229–233

Ribeiro MNG, Santos A (1975) Observações microclimaticas no ecossistema Campina Amazonica. Acta Amazonica 5:183–189

Ricci JP (1990) Les pinotières. Bois For Trop 220:55–63

Richards P, Williamson GB (1975) Treefalls and patterns of understory species in a wet lowland tropical forest. Ecology 56:1226–1229

Richards PW (1952) The tropical rain forest. An ecological study. Cambridge Univ Press, 7th edn (1979), Cambridge

Richards PW (1969) Speciation in the tropical rain forest and the concept of the niche. Biol J Linn Soc 1:149–153

Riera B (1986) A propos des chablis en forêt guyanaise Piste de St Elie. Mém Mus Natn Hist Nat Sér A 132:109–114

Rodrigues WA (1961) Estudio preliminar de mata de várzea alta de um ilha do baixo Rio Negro de sólo argiloso e úmedo. INPA (Botânica) 10:1–50

Rollet B (1974) L'architecture des forêts denses humides sempervirentes de plaine. CTFT, Nogent sur Marne

Ruddle K, Johnson D, Townsend PK, Rees JD (1978) Palm sago. A tropical starch from marginal lands. Univ Press of Hawaii, Honolulu

Rylands AB (1987) Primate communities in Amazonian forests: their habitats and food resources. Experientia 43:265–279

Sabatier D (1983) Fructification et dissémination en forêt guyanaise. L'exemple de quelques espèces ligneuses. Thesis, USTL II, Montpellier

Sabatier D (1985) Saisonnalité et déterminisme du pic de fructification en forêt guyanaise. Rev Ecol (Terre et Vie) 40:289–319

Sabatier D, Prevost MF (1989) Quelques données sur la composition floristique et la diversité des peuplements forestiers de Guyane française. Bois For Trop 219:31–55

Salazar A (1967) El aguaje (*Mauritia vinifera*) recurso forestal potential. Rev For Peru 1:65–68

Salazar A, Roessl J (1977) Estudio de potencialidad industrial del aguaje. Proyecto ITINTEC 3102 UNA-IIA, Lima

Salati E (1985) The climatology and hydrology of Amazonia. In: Prance GT, Lovejoy TE (eds) Key environments Amazonia. Pergamon Press, Oxford, pp 18–48

Salati E, Marques M (1984) Climatology on the Amazon region. In: Sioli H (ed) The Amazon limnology and landscape ecology of a mighty tropical river and its basin. Dr Junk, Dordrecht, pp 85–126

Salati E, Ribeiro MNG (1979) Floresta e clima. Acta Amazonica 9 (Suppl):15–22

Salati E, Vose PB (1984) Amazon Basin: a system in equilibrium. Science 225:129–138

Salo J, Kalliola R, Häkkinen I, Mäkinen Y, Niemela P, Puhakka M, Coley PD (1986) River dynamics and the diversity of Amazon lowland forest. Nature 322:254–258

Scariot AO, Oliveira Filho AT, Lleras E (1989) Species richness, density and distribution of palms in an eastern Amazonian seasonally flooded forest. Principes 33:172–179

Schatz GE, Williamson GB, Cogswell CM, Stam AC (1985) Stilt roots and growth of arboreal palms. Biotropica 17:206–209

Schubart HOR, Franken W, Luizão FJ (1984) Uma floresta sobre solos pobres. Ciênc Hoje 2:26–33

Schultes RE (1974) Palms and religion in the northwest Amazon. Principes 18:3–21

Schultes RE (1977) Promising structural fiber palms of the Colombian Amazon. Principes 21:72–82

Schultes RE, Raffauf RF (1990) The healing forest. Medicinal and toxic plants of the northwest Amazonia. Dioscorides Press, Portland, Oregon

Schulz JP (1960) Ecological studies on rain forest in northern Suriname. Verh K Ned Akad Wet Afd Natuurkd Tweede Reeks 53:1–267

Silva MF da, Lisbôa PLB, Lisbôa RCL (1977) Nomes vulgares de plantas amazônicas. INPA, Manaus

Simonson CH (1958) Reconnaissance soil survey of the coastal plain of British Guiana, S.A. Univ of Maryland, College Park, for International Cooperation Administration, Washington

Sioli H (1968) Hydrochemistry and geology in the Brazilian Amazonian region. Amazoniana 1–3:267–277

Sioli H (1984) The Amazon and its main affluents: Hydrography, morphology of the river sources, and river types. In: Sioli H (ed) The Amazon. Limnology and landscape ecology of a mighty tropical river and its basin. Dr Junk, Dordrecht, pp 127–165

Sioli H (1991) Introduction to the symposium: Amazonia – deforestation and possible effects. For Ecol and Manage 38:123–132

Sist P (1989a) Stratégies de régénération de quelques palmiers en forêt guyanaise. Thesis, Paris VI Univ, Paris

Sist P (1989b) Demography of *Astrocaryum sciophilum*, an understory palm of French Guiana. Principes 33:142–151

Sist P (1989c) Peuplement et phénologie des palmiers en forêt guyanaise (Piste de Saint Elie). Rev Ecol (Terre et Vie) 44:113–151

Sist P, Puig H (1987) Régénération, dynamique des populations et dissémination d'un palmier de Guyane française: *Jessenia bataua* (Mart.) Burret subsp. *oligocarpa* (Griseb. & H. Wendl.) Balick. Bull Mus Natn Hist Nat 4ème Sér 9 Sect B Adansonia 3:317–336

Skov F, Balslev H (1989) A revision of *Hyospathe* (Palmae). Nord J Bot 9:189–202

Smythe N (1978) The natural history of central American agouti (*Dasyprocta punctata*). Smithsonian contributions. Zoology 257:1–52

Smythe N, Glanz WE, Egbert G, Leigh EG Jr (1983) Population regulation in some terrestrial frugivores. In: Leigh EG Jr, Rand AS, Windsor DM (eds) The ecology of a tropical forest. Seasonal rythms and long-term changes. Smithsonian Institution Press, Washington, DC, pp 227–238

Snow DW (1981) Tropical frugivorous birds and their food plants: a word survey. Biotropica 13:1–14

Sombroek WG (1966) Amazon soils: a reconnaissance of soils of the Brazilian Amazon Region. Center of Agricultural Publication and Documentation Wageningen, Holland

Spears JS (1979) Can the wet tropical forest survive? Commonw For Rev 58:165–180

Spruce R (1871) Palmae Amazonicae, sive enumeratio palmarum in itinere suo per regiones americae aequatoriales lectorum. J Linn Soc Bot 11:65–183

Stark N, Jordan CF (1978) Nutrient retention in the root mat of an Amazonian rain forest. Ecology 59:434–437

Sternberg HO'R (1950) Vales tectônicos na planície Amazônica? Rev Bras Geogr 4:511–534

Sternberg HO'R (1955) Séismicité et morphologie en Amazonie brésilienne. Ann Géogr 342:97–105

Sternberg HO'R (1957) A propósito de meandros. Rev Bras Geogr 4:99–121

Stone TA, Brown IF, Woodwell GM (1991) Estimation, by remote sensing, of deforestation in central Rondônia, Brazil. For Ecol Manage 38:291–304

Strudwick J, Sobel GL (1988) Uses of *Euterpe oleracea* Mart. in the Amazon estuary, Brazil. Adv Econ Bot 6:225–253

Syed RA (1979) Studies on oil palm pollination by insects. Bull Entomol Res 69:213–224

Takeuchi M (1960) A estrutura da vegetacão na Amazônia I. A mata pluvial tropical. Bol Mus Para Emilio Goeldi NS Bot 6:1–17

Takeuchi M (1962) The structure of the Amazonian vegetation. VI. Igapó. J Fac Sci Univ Tokyo Sect 3 Bot 8:297–304

Teunissen PA, Wildschut JT (1970) Vegetation and flora of the savannas in the Brinckheuvel Nature Reserve, northern Suriname. North-Holland, Amsterdam London

Timm RM (1987) Tent construction by bats of the Genera *Artibeus* and *Uroderma*. In: Patterson BD, Timm RM (eds) Studies in neotropical mammalogy: essays in honor of Philip Hershkovitz, pp 187–212 (Fieldiana Zoology NS, vol 39)

Tomlinson PB (1979) Juvénilité et néoténie chez les monocotylédones. Bull Soc Bot Fr 126:227–232

Tomlinson PB (1990) The structural biology of palms. Clarendon Press, Oxford

Tostain O, Sabatier D (1988) Diversity of fruit selection in the calfbird's diet in French Guiana. Poster, OTS Silver Anniversary Symposium, University of Miami, 5–8 June 1988

Tricart J (1977) Types de lits fluviaux en Amazonie Brésilienne. Ann Géogr 473:1–54

Uhl C, Jordan CF (1984) Succession and nutrient dynamics following forest cutting and burning in Amazonia. Ecology 65:1476–1490

Uhl C, Saldarriaga J (1987) La fragilité de la forêt amazonienne. Pour la Science (Sept):38–47

Uhl NW, Dransfield J (1987) Genera Palmarum. Allen Press, Lawrence, Kansas

Ule E (1908) Die Pflanzenformationen das Amazonas-Gebietes. Bot Jahrb Syst 40:114–172, 398–443

Urrego Giraldo LE (1987) Estudio preliminar de la fenología de la canangucha (*Mauritia flexuosa* L.f.). Colombia Amazonica 2:57–81

Vandemeer JH, Stout J, Miller G (1974) Growth rates of *Welfia georgii*, *Socratea durissima* and *Iriartea gigantea* under various conditions in a natural rainforest in Costa Rica. Principes 18:148–154

Vandemeer JH, Stout J, Risch S (1979) Seed dispersal of a common Costa Rican rain forest palm (*Welfia georgii*). Trop Ecol 20:17–26

Van der Steege JG (1983) Bladproductie en Bladfytomassa van een Aantal palmsoorten wan het tropisch regenbos van Suriname. Celos Rep 138, Univ Suriname

Van Donselaar J (1965) An ecological and phytogeographic study of northern Surinam savannas. In: Lanjouw J, Versteegh PJD (eds) The vegetation of Suriname IV. Van Eedenfonds, Amsterdam

Van Donselaar J (1970a) Observations on savanna vegetation types in the Guianas. In: Lanjouw J, Bulberman FC (eds) The vegetation of Suriname VI. Van Eedenfonds, Amsterdam, pp 271–312

Van Donselaar J (1970b) Phytogeographic notes of the savanna flora of Southern Suriname. In: Lanjouw J, Bulberman FC (eds) The vegetation of Suriname VI, Van Eedenfonds, Amsterdam, pp 393–404

Van Roosmalen MGM (1980) Habitat preferences, diet, feeding strategy and social organization of the black spider monkey (*Ateles panicus panicus* Linnaeus 1758) in Surinam. RN-Rapport 80/13, Rijkqsinstitut voor Natuurbeheer Leersum

Van Steenis CGGJ (1969) Plant speciation in Malesia, with special reference to the theory of non-adaptative saltatory evolution. Biol J Linn Soc 1:97–133

Vasquez R, Gentry AH (1989) Use and misuse of forest-harvested fruits in the Iquitos area. Conserv Biol 3:1–12

Vasquez-Yanes C (1974) Studies of the germination of seeds: *Ochroma lagopus* Schwarz. Turrialba 24:176–179

Vilhena RCQ, Lins A, Luna M de S (1984) Estudo anatômico, morfológico e o crescimento das raizes adventícias de *Iriartea exorrhiza* Mart (Palmae), "Paxiúba". Bol Mus Para Emilio Goeldi NS Bot 1:87–109

Villa Nova NA, Salati E, Matsui E (1976) Estimava da evapotranspiração na Bacia Amazónica. Acta Amazonica 2:215–228

Walker I, Franken W (1983) Ecossistemas frágeis: a floresta da terra firme da Amazônia Central. Ciênc Interam 23:9–21

Wallace AR (1853) Palm trees of the Amazon and their uses. John van Voorst, London

Wessels Boer JG (1965) The indigenous palms of Suriname. Palmae. EJ Brill, Leiden

Wessels Boer JG (1968) The geonomoid palms. Verh K Ned Akad Wet Afd Natuurkd, Tweede Reeks 58:1–202

Wessels Boer JG (1988) Palmas indigenas de Venezuela. Pittieria 17 Merida

Wheeler MA (1970) Siona use of chambira palm fiber. Econ Bot 24:180–181

Whitmore TC (1978) Gaps in the forest canopy. In: Tomlinson PB, Zimmermann MH (eds) Tropical trees as living systems. Cambridge Univ Press, Cambridge, pp 639–655

Whitmore TC (1982) On pattern and process in forests. In: Newman EI (ed) The plant community as a working mechanism. Blackwell, Oxford, pp 45–59

Whitmore TC (1983) Secondary succession from seeds in tropical rain forests. For Abstr 44:767–779

Wilbert J (1980) The palm leaf-sail of the Warao Indians. Principes 24:162–169

Yeaton RI (1979) Intraspecific competition in a population of the stilt palm, *Socratea durissima* (Oerst). Wendl. on Barro Colorado Island, Panama. Biotropica 11:155–158

Zona S, Henderson A (1989) A review of animal-mediated seed dispersal of palms. Selbyana 11:6–21

Index of Latin Names

Bold Face indicates elementary descriptions of genera and illustrations.

The following names are provisional: *Astrocaryum carnosum, A. gratum,* and *A. scopatum*. They will be validly published in the sense of the International Code of Botanical Nomenclature in: Kahn F, Millan B (1992) *Astrocaryum* in Amazonia. Bull Inst Fr Etudes Andines 21 (2)

Subject Index

Bold Face indicates illustrations.